TSA Practice Papers

Volumes One & Two

UniAdmissions

Published by *RAR Medical Services Limited*
www.uniadmissions.co.uk
info@uniadmissions.co.uk
Tel: 0208 068 0438

TSA Practice Papers

6 Full Papers & Solutions

Rohan Agarwal
Jonathan Madigan

UniAdmissions

About the Authors

Jon studied Economics and Management at St Hugh's College, Oxford, between 2013 and 2016. He sat the Thinkings Skills Assessment and **scored full marks** in section 1 of the paper, placing him in the **top 0.1% of candidates** who sat the assessment that year.

Jon has worked with UniAdmissions since 2014, working primarily as a personal tutor for a number of applicants across the entire application process. He has also provided Thinking Skills Assessment preparation courses within schools, and remains very familiar with all aspects of the paper.

Although Jon has moved into finance since graduation, he remains involved with tutoring as much as time allows. He regularly visits Oxford in his free time, and hopes to return for further study at some point in the future.

Rohan is the **Director of Operations** at *UniAdmissions* and is responsible for its technical and commercial arms. He graduated from Gonville and Caius College, Cambridge and is a fully qualified doctor. Over the last five years, he has tutored hundreds of successful Oxbridge and Medical applicants. He has also authored ten books on admissions tests and interviews.

Rohan has taught physiology to undergraduates and interviewed medical school applicants for Cambridge. He has published research on bone physiology and writes education articles for the Independent and Huffington Post. In his spare time, Rohan enjoys playing the piano and table tennis.

Introduction

The Basics

The Thinking Skills Assessment is an aptitude test taken by students who are applying to certain courses at Cambridge and Oxford. Cambridge applicants sit the TSA Cambridge and Oxford applicants sit the TSA Oxford.

SECTION	SKILLS TESTED	QUESTIONS	TIMING
ONE	Problem-solving skills, including numerical and spatial reasoning. Critical thinking skills, including understanding argument and reasoning using everyday language.	50 MCQs	90 minutes
TWO	Ability to organise ideas in a clear and concise manner, and communicate them effectively in writing. Questions are usually but not necessarily medical.	One Essay from Four	30 minutes

NB: **TSA Oxford** consists of sections 1 + 2; **TSA Cambridge** and **TSA UCL** consist of only section 1.

Who has to sit the TSA?

Exam	Course
TSA Oxford	Students applying for the following subjects at Oxford **MUST** take the TSA: Economics and Management Experimental Psychology History and Economics Human Sciences Philosophy and Linguistics Philosophy, Politics and Economics (PPE) Psychology and Linguistics Psychology and Philosophy
TSA Cambridge	Students applying for Land Economy **MAY** have to take the TSA depending on the college they apply to.
TSA UCL	Students applying for European Social and Political Students **MUST** take the TSA.

NB: Applicants for Oxford **Chemistry or History & Economics only have to complete section 1 of the TSA.**

Preparing for the TSA

Before going any further, it's important that you understand the optimal way to prepare for the TSA. Rather than jumping straight into doing mock papers, it's essential that you start by understanding the components and the theory behind the TSA by using a TSA textbook. Once you've finished the non-timed practice questions, you can progress to past TSA papers. These are freely available online at www.uniadmissions.co.uk/tsa-past-papers and serve as excellent practice. You're strongly advised to use these in combination with the *TSA Past Worked Solutions* Book so that you can improve your weaknesses. Finally, once you've exhausted past papers, move onto the mock papers in this book.

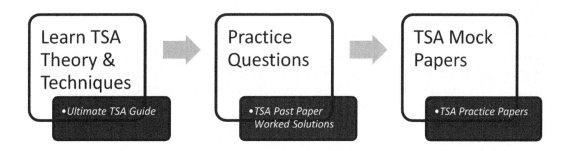

Already seen them all?

So, you've run out of past papers? Well hopefully that is where this book comes in. It contains six unique mock papers; each compiled by TSA Expert tutors at *UniAdmissions* and available nowhere else.

Having successfully gained a place at Oxford and scoring in the top 10% of the TSA, our tutors are intimately familiar with the TSA and its associated admission procedures. So, the novel questions presented to you here are of the correct style and difficulty to continue your revision and stretch you to meet the demands of the TSA.

General Advice

Start Early

It is much easier to prepare if you practice little and often. Start your preparation well in advance; ideally 10 weeks but at the latest within a month. This way you will have plenty of time to complete as many papers as you wish to feel comfortable and won't have to panic and cram just before the test, which is a much less effective and more stressful way to learn. In general, an early start will give you the opportunity to identify the complex issues and work at your own pace.

Prioritise

The MCQ section can be very time-pressured, and if you fail to answer the questions within the time limit you will be doing yourself a major disservice as every mark counts for this section. You need to be aware of how much time you're spending on each passage and allocate your time wisely. For example, since there are 50 questions in Section 1, and you are given 90 minutes in total, you will ideally take about 100 seconds per question (including reading time) so that you will not run out of time and panic towards the end.

Positive Marking

There are no penalties for incorrect answers; you will gain one for each right answer and will not get one for each wrong or unanswered one. This provides you with the luxury that you can always guess should you absolutely be not able to figure out the right answer for a question or run behind time. Since each question in Section 1 provides you with 5 possible answers, you have a 20% chance of guessing correctly. Therefore, if you aren't sure (and are running short of time), then make an educated guess and move on. Before 'guessing' you should try to eliminate a couple of answers to increase your chances of getting the question correct. For example, if a question has 5 options and you manage to eliminate 2 options- your chances of getting the question increase from 20% to 33%!

Avoid losing easy marks on other questions because of poor exam technique. Similarly, if you have failed to finish the exam, take the last ten seconds to guess the remaining questions to at least give yourself a chance of getting them right.

Practice

This is the best way of familiarising yourself with the style of questions and the timing for this section. Although the test does not demand any technical knowledge, you are unlikely to be familiar with the style of questions in all sections when you first encounter them. Therefore, you want to be comfortable at using this before you sit the test.

Practising questions will put you at ease and make you more comfortable with the exam. The more comfortable you are, the less you will panic on the test day and the more likely you are to score highly. Initially, work through the questions at your own pace, and spend time carefully reading the questions and looking at any additional data. When it becomes closer to the test, **make sure you practice the questions under exam conditions**.

Past Papers

Official past papers and answers are freely available online at **www.uniadmissions.co.uk/tsa-past-papers**. Practice makes perfect, and the more you practice the questions, especially for Section 1, the better you will get. Do not worry if you make plenty of mistakes at the start, the best way to learn is to understand why you have made certain mistakes and to not commit them again in the future!

You will undoubtedly get stuck when doing some past paper questions – they are designed to be tricky and the answer schemes don't offer any explanations. Thus, **you're highly advised to acquire a copy of *TSA Past Paper Worked Solutions*** – a free ebook is available online (see the back of this book for more details).

Repeat Questions

When checking through answers, pay particular attention to questions you have got wrong. If there is a worked answer, look through that carefully until you feel confident that you understand the reasoning, and then repeat the question without help to check that you can do it. If only the answer is given, have another look at the question and try to work out why that answer is correct. This is the best way to learn from your mistakes, and means you are less likely to make similar mistakes when it comes to the test. The same applies for questions which you were unsure of and made an educated guess which was correct, even if you got it right. When working through this book, **make sure you highlight any questions you are unsure of**, this means you know to spend more time looking over them once marked.

No Calculators and Dictionaries

The TSA requires a strong command of the English language, especially for Section 2 where you are asked to write an essay in 30 minutes. You are not allowed to use spell check or a dictionary, hence you should ensure that you written English is up to standard and you should ideally make close to no grammatical or spelling errors for your essay.

Section 1 contains several numerical reasoning questions, and you are not allowed to use a calculator, so make sure you are careful with your calculations.

> ***Top tip!*** Section 1 is arguably the more important Section, even for universities that ask for both sections to be completed – make sure you do enough practices to improve your score.

Keywords

If you're stuck on a question, sometimes you can simply quickly scan the passage for any keywords that match the questions.

A word on Timing...

"If you had all day to do your exam, you would get 100%. But you don't."

Whilst this isn't completely true, it illustrates a very important point. Once you've practiced and know how to answer the questions, the clock is your biggest enemy. This seemingly obvious statement has one very important consequence. **The way to improve your score is to improve your speed.** There is no magic bullet. But there are a great number of techniques that, with practice, will give you significant time gains, allowing you to answer more questions and score more marks.

Timing is tight throughout – **mastering timing is the first key to success**. Some candidates choose to work as quickly as possible to save up time at the end to check back, but this is generally not the best way to do it. Often questions can have a lot of information in them – each time you start answering a question it takes time to get familiar with the instructions and information. By splitting the question into two sessions (the first run-through and the return-to-check) you double the amount of time you spend on familiarising yourself with the data, as you have to do it twice instead of only once. This costs valuable time. In addition, candidates who do check back may spend 2–3 minutes doing so and yet not make any actual changes. Whilst this can be reassuring, it is a false reassurance as it is unlikely to have a significant effect on your actual score. Therefore, it is usually best to pace yourself very steadily, aiming to spend the same amount of time on each question and finish the final question in a section just as time runs out. This reduces the time spent on re-familiarising with questions and maximises the time spent on the first attempt, gaining more marks.

It is essential that you don't get stuck with the hardest questions – no doubt there will be some. In the time spent answering only one of these you may miss out on answering three easier questions. If a question is taking too long, choose a sensible answer and move on. Never see this as giving up or in any way failing, rather it is the smart way to approach a test with a tight time limit. With practice and discipline, you can get very good at this and learn to maximise your efficiency. It is not about being a hero and aiming for full marks – this is almost impossible and very much unnecessary. It is about maximising your efficiency and gaining the maximum possible number of marks within the time you have.

Use the Options:

Some passages may try to trick you by providing a lot of unnecessary information. When presented with long passages that are seemingly hard to understand, it's essential you look at the answer options so you can focus your mind. This can allow you to reach the correct answer a lot more quickly. Consider the example below:

'Mountain climbing is viewed by some as an extreme sport, while for others it is simply an exhilarating pastime that offers the ultimate challenge of strength, endurance, and sacrifice. It can be highly dangerous, even fatal, especially when the climber is out of his or her depth, or simply gets overwhelmed by weather, terrain, ice, or other dangers of the mountain. Inexperience, poor planning, and inadequate equipment can all contribute to injury or death, so knowing what to do right matters.

Despite all the negatives, when done right, mountain climbing is an exciting, exhilarating, and rewarding experience. This article is an overview beginner's guide and outlines the initial basics to learn. Each step is deserving of an article in its own right, and entire tomes have been written on climbing mountains, so you're advised to spend a good deal of your beginner's learning immersed in reading widely. This basic overview will give you an idea of what is involved in a climb.'

Looking at the options first makes it obvious that certain information is redundant and allows you to quickly zoom in on certain keywords you should pick up on in order to answer the questions.

In other cases, **you may actually be able to solve the question without having to read the passage over and over again**. For example:

Which statement best summarises this paragraph?
A. *Mountain climbing is an extreme sport fraught with dangers.*
B. *Without extensive experience embarking on a mountain climb is fatal.*
C. *A comprehensive literature search is the key to enjoying mountain climbing.*
D. *Mountain climbing is difficult and is a skill that matures with age if pursued.*
E. *The terrain is the biggest unknown when climbing a mountain and therefore presents the biggest danger.*

If you read the passage first before looking at the question, you might have forgotten what the passage mentioned, and you will have to spend extra time going back to the passage to re-read it again.

You can **save a lot of time by looking at the questions first before reading the passage**. After looking at the question, you will know at the back of your head to look out for and this will save a considerable amount of time.

Manage your Time:

It is highly likely that you will be juggling your revision alongside your normal school studies. Whilst it is tempting to put your A-levels on the back burner falling behind in your school subjects is not a good idea, don't forget that to meet the conditions of your offer should you get one you will need at least one A*. So, time management is key!

Make sure you set aside a dedicated 90 minutes (and much more closer to the exam) to commit to your revision each day. The key here is not to sacrifice too many of your extracurricular activities, everybody needs some down time, but instead to be efficient. Take a look at our list of top tips for increasing revision efficiency below:

1. Create a comfortable work station
2. Declutter and stay tidy
3. Treat yourself to some nice stationery
4. See if music works for you → if not, find somewhere peaceful and quiet to work
5. Turn off your mobile or at least put it into silent mode
6. Silence social media alerts
7. Keep the TV off and out of sight
8. Stay organised with to do lists and revision timetables – more importantly, stick to them!
9. Keep to your set study times and don't bite off more than you can chew
10. Study while you're commuting
11. Adopt a positive mental attitude
12. Get into a routine
13. Consider forming a study group to focus on the harder exam concepts
14. Plan rest and reward days into your timetable – these are excellent incentive for you to stay on track with your study plans!

Keep Fit & Eat Well:

'A car won't work if you fill it with the wrong fuel' - your body is exactly the same. You cannot hope to perform unless you remain fit and well. The best way to do this is not underestimate the importance of healthy eating. Beige, starchy foods will make you sluggish; instead start the day with a hearty breakfast like porridge. Aim for the recommended 'five a day' intake of fruit/veg and stock up on the oily fish or blueberries – the so called "super foods".

When hitting the books, it's essential to keep your brain hydrated. If you get dehydrated you'll find yourself lethargic and possibly developing a headache, neither of which will do any favours for your revision. Invest in a good water bottle that you know the total volume of and keep sipping throughout the day. Don't forget that the amount of water you should be aiming to drink varies depending on your mass, so calculate your own personal recommended intake as follows: 30 ml per kg per day.

It is well known that exercise boosts your wellbeing and instils a sense of discipline. All of which will reflect well in your revision. It's well worth devoting half an hour a day to some exercise, get your heart rate up, break a sweat, and get those endorphins flowing.

Sleep

It's no secret that when revising you need to keep well rested. Don't be tempted to stay up late revising as sleep actually plays an important part in consolidating long term memory. Instead aim for a minimum of 7 hours good sleep each night, in a dark room without any glow from electronic appliances. Install flux (https://justgetflux.com) on your laptop to prevent your computer from disrupting your circadian rhythm. Aim to go to bed the same time each night and no hitting snooze on the alarm clock in the morning!

Revision Timetable

Still struggling to get organised? Then try filling in the example revision timetable below, remember to factor in enough time for short breaks, and stick to it! Remember to schedule in several breaks throughout the day and actually use them to do something you enjoy e.g. TV, reading, YouTube etc.

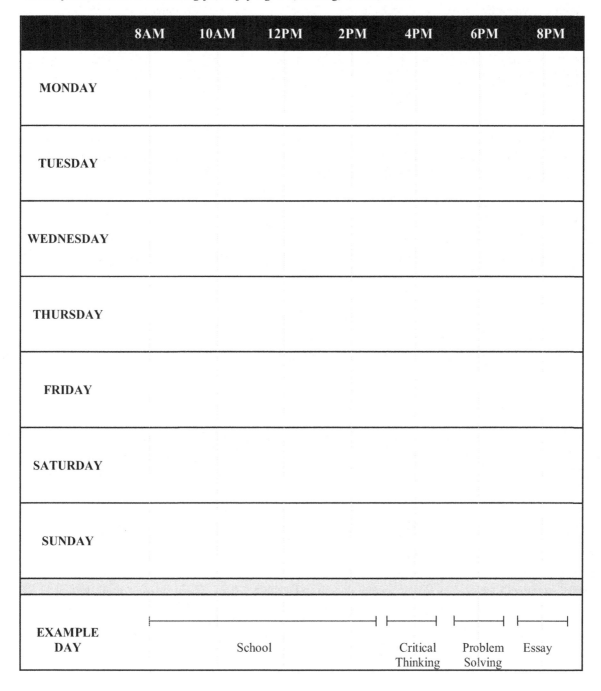

	8AM	10AM	12PM	2PM	4PM	6PM	8PM
MONDAY							
TUESDAY							
WEDNESDAY							
THURSDAY							
FRIDAY							
SATURDAY							
SUNDAY							
EXAMPLE DAY	School				Critical Thinking	Problem Solving	Essay

Top tip! Ensure that you take a watch that can show you the time in seconds into the exam. This will allow you have a much more accurate idea of the time you're spending on a question. In general, if you've spent >150 seconds on a section 1 question – move on regardless of how close you think you are to solving it.

Getting the most out of Mock Papers

Mock exams can prove invaluable if tackled correctly. Not only do they encourage you to start revision earlier, they also allow you to **practice and perfect your revision technique**. They are often the best way of improving your knowledge base or reinforcing what you have learnt. Probably the best reason for attempting mock papers is to familiarise yourself with the exam conditions of the TSA as they are particularly tough.

Start Revision Earlier

Thirty five percent of students agree that they procrastinate to a degree that is detrimental to their exam performance. This is partly explained by the fact that they often seem a long way in the future. In the scientific literature this is well recognised, Dr. Piers Steel, an expert on the field of motivation states that *'the further away an event is, the less impact it has on your decisions'*.

Mock exams are therefore a way of giving you a target to work towards and motivate you in the run up to the real thing – every time you do one treat it as the real deal! If you do well then it's a reassuring sign; if you do poorly then it will motivate you to work harder (and earlier!).

Practice and perfect revision techniques

In case you haven't realised already, revision is a skill all to itself, and can take some time to learn. For example, the most common revision techniques including **highlighting and/or re-reading are quite ineffective** ways of committing things to memory. Unless you are thinking critically about something you are much less likely to remember it or indeed understand it.

Mock exams, therefore allow you to test your revision strategies as you go along. Try spacing out your revision sessions so you have time to forget what you have learnt in-between. This may sound counterintuitive but the second time you remember it for longer. Try teaching another student what you have learnt; this forces you to structure the information in a logical way that may aid memory. Always try to question what you have learnt and appraise its validity. Not only does this aid memory but it is also a useful skill for the TSA, Oxbridge interviews, and beyond.

Improve your knowledge

The act of applying what you have learnt reinforces that piece of knowledge. An essay question in Section 2 may ask you about a fairly simple topic, but if you have a deep understanding of it you are able to write a critical essay that stands out from the crowd. Essay questions in particular provide a lot of room for students who have done their research to stand out, hence you should always aim to improve your knowledge and apply it from time to time. As you go through the mocks or past papers take note of your performance and see if you consistently under-perform in specific areas, thus highlighting areas for future study.

Get familiar with exam conditions

Pressure can cause all sorts of trouble for even the most brilliant students. The TSA is a particularly time pressured exam with high stakes – your future (without exaggerating) does depend on your result to a great extent. The real key to the TSA is overcoming this pressure and remaining calm to allow you to think efficiently.

Mock exams are therefore an excellent opportunity to devise and perfect your own exam techniques to beat the pressure and meet the demands of the exam. **Don't treat mock exams like practice questions – it's imperative you do them under time conditions.**

Remember! It's better that you make all the mistakes you possibly can now in mock papers and then learn from them so as not to repeat them in the real exam.

Before using this Book

Do the ground work

➢ Understand the format of the TSA – have a look at the TSA website and familiarise yourself with it: www.admissionstesting.org/thinking-skills-assessment

➢ Read widely in order to prepare yourself for the essay.

➢ Improve your written English if you are not confident in this aspect by practicing writing and reading frequently.

➢ Try to broaden your reading by learning about different topics that you are unfamiliar with as the essay topics can vary greatly.

➢ Learn how to understand a writer's viewpoint by reading news articles and having a go at summarising what the writer is arguing about.

➢ Be consistent – slot in regular TSA practice sessions when you have pockets of free time.

➢ Engage in discussion sessions with your friends and teachers – this might give you more ideas about certain essay topics.

Ease in gently

With the ground work laid, there's still no point in adopting exam conditions straight away. Instead invest in a beginner's guide to the TSA, which will not only describe in detail the background and theory of the exam, but take you through section by section what is expected. *The Ultimate TSA Guide* is the most popular TSA textbook – you can get a free copy by flicking to the back of this book.

When you are ready to move on to past papers, take your time and puzzle your way through all the questions. Really try to understand solutions. A past paper question won't be repeated in your real exam, so don't rote learn methods or facts. Instead, focus on applying prior knowledge to formulate your own approach.

If you're really struggling and have to take a sneak peek at the answers, then practice thinking of alternative solutions, or arguments for essays. It is unlikely that your answer will be more elegant or succinct than the model answer, but it is still a good task for encouraging creativity with your thinking. Get used to thinking outside the box!

Accelerate and Intensify

Start adopting exam conditions after you've done two past papers. Don't forget that **it's the time pressure that makes the TSA hard** – if you had as long as you wanted to sit the exam you would probably get 100%. If you're struggling to find comprehensive answers to past papers then *TSA Past Papers Worked Solutions* contains detailed explained answers to every TSA past paper question and essay (flick to the back to get a free copy).

Doing all the past papers is a good target for your revision. Choose a paper and proceed with strict exam conditions. Take a short break and then mark your answers before reviewing your progress. For revision purposes, as you go along, keep track of those questions that you guess – these are equally as important to review as those you get wrong.

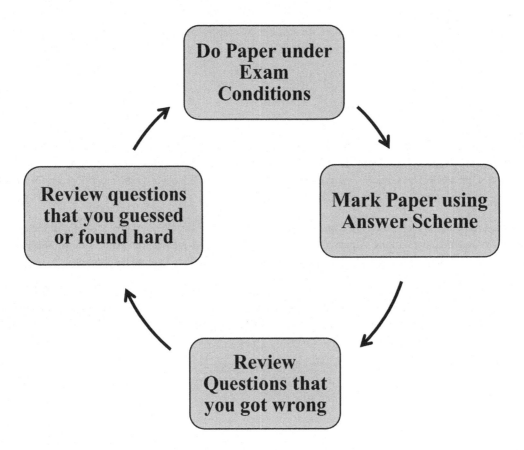

Once you've exhausted all the past papers, move on to tackling the unique mock papers in this book. In general, you should aim to complete one to two mock papers every night in the ten days preceding your exam.

Section 1: An Overview

What will you be tested on?	No. of Questions	Duration
Problem-solving skills, numerical and spatial reasoning, critical thinking skills, understanding arguments and reasoning	50 MCQs	90 Minutes

This is the first section of the TSA, comprising a total of 50 MCQ questions. You have 90 minutes in total to complete the MCQ questions, including reading time. In order to keep within the time limit, you realistically have about 108 seconds per question as you have to factor in the reading time as well.

Not all the questions are of equal difficulty and so as you work through the past material it is certainly worth learning to recognise quickly which questions you should spend less time on in order to give yourself more time for the trickier questions.

Deducing arguments

Several MCQ questions will be aimed at testing your understand of the writer's argument. It is common to see questions asking you 'what is the writer's view?' or 'what is the writer trying to argue?'. This is arguably an important skill you will have to develop, and the TSA is designed to test this ability. You have limited time to read the passage and understand the writer's argument, and the only way to improve your reading comprehension skill is to read several well-written news articles on a daily basis and think about them in a critical manner.

Assumptions

It is important to be able to identify the assumptions that a writer makes in the passage, as several questions might question your understand of what is assumed in the passage. For example, if a writer mentions that 'if all else remains the same, we can expect our economic growth to improve next year', you can identify an assumption being made here – the writer is clearly assuming that all external factors remain the same.

Fact vs. Opinion

It is important to **be able to decipher whether the writer is stating a fact or an opinion** – the distinction is usually rather subtle and you will have to decide whether the writer is giving his or her own personal opinion, or presenting something as a fact. Section 1 may contain questions that will test your ability to identify what is presented as a fact and what is presented as an opinion.

Fact	Opinion
'There are 7 billion people in this world...'	'I believe there are more than 7 billion people in this world...'
'She is an Australian...'	'She sounded like an Australian...'
'Trump is the current President...'	'Trump is a horrible President...'
'Vegetables contain a lot of fibre...'	'Vegetables are good for you...'

Numerical and spatial reasoning

There are several questions that will test how well you can cope with numbers, and you should ideally be comfortable with simple mental calculations and being able to think logically.

Section 2: An Overview

What will you be tested on?	No. of Questions	Duration
Your ability to write an essay under timed conditions, your writing technique and your argumentative abilities	1 out of 4	30 Minutes

Section 2 is usually what students are more comfortable with – after all, many GCSE and A Level subjects require you to write essays within timed conditions. This section does not require you to have any particular specialist knowledge – the questions can be very broad and cover a wide range of topics.

Here are some of the topics that might appear in Section 2:

- ➢ Science
- ➢ Politics
- ➢ Religion
- ➢ Technology

- ➢ Ethics
- ➢ Morality
- ➢ Philosophy
- ➢ Education

- ➢ History
- ➢ Geopolitics

As you can see, this list is very broad and definitely non-exhaustive, and you do not get many choices to choose from (you have to write one essay out of three choices). Many students make the mistake of focusing too narrowly on one or two topics that they are comfortable with – this is a dangerous gamble and if you end up with four questions you are unfamiliar with, this is likely to negatively impact your score. **You should ideally focus on three topics to prepare from the list above**, and you can pick and choose which topics from the list above are the ones you would be more interested in. Here are some suggestions:

Science
An essay that is related to science might relate to recent technological advancements and their implications, such as the rise of Bitcoin and the use of blockchain technology and artificial intelligence. This is interrelated to ethical and moral issues, hence you cannot merely just regurgitate what you know about artificial intelligence or blockchain technology. **The examiners do not expect you to be an expert in an area of science** – what they want to see is how you identify certain moral or ethical issues that might arise due to scientific advancements, and how do we resolve such conundrums as human beings.

Politics
Politics is undeniably always a hot topic and consequently a popular choice amongst students. The danger with writing a politics question is that some **students get carried away and make their essay too one-sided or emotive** – for example a student may chance upon an essay question related to Brexit and go on a long rant about why the referendum was a bad idea. You should always remember to answer the question and make sure your essay addresses the exact question asked – do not get carried away and end up writing something irrelevant just because you have strong feelings about a certain topic.

Religion
Religion is always a controversial issue and essays on religion provide good students with an excellent opportunity to stand out and display their maturity in thought. Questions can range from asking about your opinion with regards to banning the wearing of a headdress to whether children should be exposed to religious practices at a young age. Questions related to religion will **require you to be sensitive and measured** in your answers - it is easy to trip up on such questions if you're not careful.

Education
Education is perhaps always a relatable topic to students, and students can draw from their own experience with the education system in order to form their opinion and write good essays on such topics. Questions can range from whether university places should be reduced, to whether we should be focusing on learning the sciences as opposed to the arts.

Section 2: Revision Guide

Science

Resource	What to read/do
1. Newspaper Articles	• The Guardian, The Times, The Economist, The Financial Times, The Telegraph, The New York Times, The Independent
2. A Levels/IB	• Look at the content of your science A Levels/IB if you are doing science subjects and critically analyse what are the potential moral/ethical implications • Use your A Levels/IB resources in order to seek out further readings – e.g. links to a scientific journal or blog commentary • Remember that for your LNAT essay you should not focus on the technical issues too much – think more about the ethical and moral issues
3. Online videos	• There are plenty of free resources online that provide interesting commentary on science and the moral and ethical conundrums that scientists face on a daily basis • E.g. Documentaries and specialist science channels on YouTube • National Geographic, Animal Planet etc. might also be good if you have access to them
4. Debates	• Having a discussion with your friends about topics related to science might also help you formulate some ideas • Attending debate sessions where the topic is related to science might also provide you with excellent arguments and counter-arguments • Some universities might also host information sessions for sixth form students – some might be relevant to ethical and moral issues in science
5. Museums	• Certain museums such as the Natural Science Museum might provide some interesting information that you might not have known about
6. Non-fiction books	• There are plenty of non-fiction books (non-technical ones) that might discuss moral and ethical issues about science in an easily digestible way

Politics

Resource	What to read/do
1. Newspaper Articles	• The Guardian, The Times, The Economist, The Financial Times, The Telegraph, The New York Times, The Independent
2. Television	• Parliamentary sessions • Prime Minister Questions • Political news
3. Online videos	• Documentaries • YouTube Channels
4. Lectures	• University introductory lectures • Sixth form information sessions
5. Debates	• Debates held in school • Joining a politics club
6. Podcasts	• Political podcasts • Listen to both sides to get a more rounded view (e.g. listening to both left and right wing podcasts)

Religion

Syllabus Point	What to read/do
1. **Newspaper Articles**	• The Guardian, The Times, The Economist, The Financial Times, The Telegraph, The New York Times, The Independent
2. **Non-fiction books**	• Read up about books that explain the origins and beliefs of different types of religion • E.g. Books that talk about the origins of Christianity, Islam or Buddhism, theology books etc.
3. **Talking to religious leaders**	• Talking to religious leaders may be a good way of understanding different religions more and being able to write an essay on religion with more maturity and nuance • Talking to people from different religious backgrounds may also be a good way of forming a more well-rounded opinion
4. **Online videos**	• Documentaries on religion • YouTube channels providing informative and educational videos on different religions – e.g. history, background
5. **Lectures**	• Information sessions • Relevant introductory lectures
6. **Opinion articles**	• Informative blogs and journals • Read both arguments and counter-arguments and come up with your own viewpoint

Education

Syllabus Point	What to read/do
1. **Newspaper Articles**	• The Guardian, The Times, The Economist, The Financial Times, The Telegraph, The New York Times, The Independent
2. **A Levels/IB**	• Draw inspiration from what you are studying in your A Levels or IB – do you feel like what you are studying is useful and relevant? E.g. Studying arts versus science • Compare the education you are receiving with your friends in different schools or different subjects
3. **Educational exchange**	• If you have an opportunity to go on an educational exchange, this might be a good opportunity to compare and contrast different educational systems • E.g. the approach to education in Germany versus the UK
4. **University applications**	• Have a read of how different universities promote themselves – do they claim to provide students with academic enlightenment, or better job prospects, or a good social life? • Why do different universities focus on different things?
5. **Online videos**	• Documentaries • YouTube Channels
6. **Talk to your teachers**	• Your teachers have been in the education industry for years and maybe decades – talk to them and ask them for their opinion • Talk to different teachers and compare their opinions regarding how we should approach education

---

Top Tip! Although you aren't required to have extra knowledge for the TSA essay, doing so will allow you to make your essay stand out from the crowd. However, you should first prioritise perfecting your writing style rather than doing extra reading as the former will have a greater impact on your mark.

How to use this Book

If you have done everything this book has described so far then you should be well equipped to meet the demands of the TSA, and therefore **the mock papers in the rest of this book should ONLY be completed under exam conditions**.

This means:
➢ Absolute silence – no TV or music
➢ Absolute focus – no distractions such as eating your dinner
➢ Strict time constraints – no pausing half way through
➢ No checking the answers as you go
➢ Give yourself a maximum of three minutes between sections – keep the pressure up
➢ Complete the entire paper before marking
➢ Mark harshly

In practice this means setting aside 90 minutes for Section 1 and 30 minutes for Section 2 in an evening to find a quiet spot without interruptions and tackle the paper. Completing one mock paper every evening in the week running up to the exam would be an ideal target.

➢ Tackle the paper as you would in the exam.
➢ Return to mark your answers, but mark harshly if there's any ambiguity.
➢ Highlight any areas of concern.
➢ If warranted read up on the areas you felt you underperformed to reinforce your knowledge.
➢ If you inadvertently learnt anything new by muddling through a question, go and tell somebody about it to reinforce what you've discovered.

Finally relax… the TSA is an exhausting exam, concentrating so hard continually for 1.5 hours will take its toll. So, being able to relax and switch off is essential to keep yourself sharp for exam day! Make sure you reward yourself after you finish marking your exam.

Top Tip! You can get free copies of *The Ultimate TSA Guide* and *TSA Past Paper Worked Solutions* books by flicking to the back of this book.

Scoring Tables

Use these to keep a record of your scores from past papers – you can then easily see which paper you should attempt next (always the one with the lowest score).

SECTION 1	1st Attempt	2nd Attempt	3rd Attempt
2008			
2009			
2010			
2011			
2012			
2013			
2014			
2015			
2016			
2017			

SECTION 1	1st Attempt	2nd Attempt	3rd Attempt
Mock Paper A			
Mock Paper B			
Mock Paper C			
Mock Paper D			
Mock Paper E			
Mock Paper F			

You will not be able to give yourself a score for Section 2– the best way to gauge your performance for Section 2 will be to compare your arguments and counter-arguments with the model answer, or let your friends or teachers read it and gather some feedback from them. Fortunately for the mock papers in this book, there are model answers for you to compare your essays against!

Top Tip! When repeating a mock paper, its best to attempt a different essay title to give yourself maximum experience with the various styles of TSA essays.

Mock Papers

Mock Paper A

Question 1:

"Competitors need to be able to run 200 metres in under 25 seconds to qualify for a tournament. James, Steven and Joe are attempting to qualify. Steven and Joe run faster than James. James' best time over 200 metres is 26.2 seconds."

Which response is **definitely** true?

A. Only Joe qualifies.   D. Joe qualifies.
B. James does not qualify.   E. No one qualifies.
C. Joe and Steven both qualify.

Question 2:

You spend £5.60 in total on a sandwich, a packet of crisps and a watermelon. The watermelon cost twice as much as the sandwich, and the sandwich cost twice the price of the crisps.

How much did the watermelon cost?

A. £1.20   B. £2.60   C. £2.80   D. £3.20   E. £3.60

Question 3:

Jane, Chloe and Sam are all going by train to a football match. Chloe gets the 2:15pm train. Sam's journey takes twice as long Jane's. Sam catches the 3:00pm train. Jane leaves 20 minutes after Chloe and arrives at 3:25pm.

When will Sam arrive?

A. 3:50pm   B. 4:10pm   C. 4:15pm   D. 4:30pm   E. 4:40pm

Question 4:

Michael has eleven sweets. He gives three sweets to Hannah. Hannah now has twice the number of sweets Michael has remaining. How many sweets did Hannah have before the transaction?

A. 11   B. 12   C. 13   D. 14   E. 1

Question 5:

"Alex's current weekly take-home pay is £250 per week. Alex is to receive a pay rise of 5% plus an extra £6 per week. The flat rate of income tax will decrease from 14% to 12% at the same time."

What will his new weekly take-home pay be, to the nearest whole pound?

A. £260   B. £267   C. £274   D. £279   E. £285

Question 6:

"You have four boxes, each containing two coloured cubes. Box A contains two white cubes, Box B contains two black cubes, and Boxes C and D both contain one white cube and one black cube. You pick a box at random and take out one cube. It is a white cube. You then draw another cube from the same box."

What is the probability that this cube is not white?

A. ½   B. ⅓   C. ⅔   D. ¼   E. ¾

Question 7:

"Anderson & Co. hire out heavy plant machinery at a cost of £500 per day. There is a surcharge for heavy usage, at a rate of £10 per minute of usage over 80 minutes. Concordia & Co. charge £600 per day for similar machinery, plus £5 for every minute of usage."

At what duration of usage are the costs the same for both companies?

A. 100 minutes
B. 130 minutes
C. 140 minutes

D. 170 minutes
E. 180 minutes

Question 8:

"Simon is discussing with Seth whether or not a candidate is suitable for a job. When pressed for a weakness at interview, the candidate told Simon that he is a slow eater. Simon argues that this will reduce the candidate's productivity, since he will be inclined to take longer lunch breaks."

*Which statement **best** substantiates Simon's argument?*

A. Slow eaters will take longer to eat lunch
B. Longer lunch breaks are a distraction
C. Eating more slowly will reduce the time available to work

D. Eating slowly is a weakness
E. People who like food are more likely to eat slowly

Question 9:

Three pieces of music are on repeat in different rooms of a house. One piece of music is three minutes long, one is four minutes long and the final one is 100 seconds long. All pieces of music start playing at exactly the same time.

How long is it until they are next all starting together?

A. 12 minutes
B. 15 minutes
C. 20 minutes

D. 60 minutes
E. 300 minutes

Question 10:

A car leaves Salisbury at 8:22am and travels 180 miles to Lincoln, arriving at 12:07pm. Near Warwick, the driver stopped for a 14 minute break.

What was its average speed, whilst travelling, in kilometres per hour? It should be assumed that the conversion from miles to kilometres is 1:1.6.

A. 51kph   B. 67kph   C. 77kph   D. 82kph   E. 86kph

Questions **11** and **12** refer to the following data:

Five respondents were asked to estimate the value of three bottles of wine, in pounds sterling.

Respondent	Wine 1	Wine 2	Wine 3
1	13	16	25
2	17	16	23
3	11	17	21
4	13	15	14
5	15	19	29
Actual retail value	8	25	23

Question 11:

What is the mean error margin in the guessing of the value of wine 1?

A. £4.80   B. £5.60   C. £5.80   D. £6.20   E. £6.40

Question 12:

Which respondent guessed most accurately on average?

A. Respondent 1   D. Respondent 4
B. Respondent 2   E. Respondent 5
C. Respondent 3

Questions **13** and **14** refer to the following data

The population of Country A is 40% greater than the population of Country B.

The population of Country C is 30% less than the population of Country D (which has a population 20% greater than Country B).

Question 13:

Given that the population of Country A is 45 million, what is the population of country D?

A. 32.1 million   D. 38.5 million
B. 35.8 million   E. 39.0 million
C. 36.6 million

Question 14:

The population of Country A is still 45 million. If Country B introduced a new health initiative costing $45 per capita, what would be the total cost?

A. $1.34 billion   D. $1.56 billion
B. $1.44 billion   E. $1.66 billion
C. $1.50 billion

Question 15:

A car averages a speed of 30mph over a certain distance and then returns over the same distance at an average speed of 20mph.

What is the average speed for the journey as a whole?

A. 22.5 mph   D. 26 mph
B. 24 mph   E. The distance travelled is required to calculate
C. 25 mph   average speed

Question 16:

"All sheep are ruminants and all marsupials are mammals. No sheep are marsupials."

Which of the following must be true?

A. Some ruminants are marsupials.

B. All mammals are marsupials.

C. All sheep are mammals.

D. Some sheep are marsupials.

E. None of the above

Question 17:

The price of toothpaste rises by 80%. This is later reduced by 50% due to competition. Zoe buys two tubes of toothpaste and gets the third free because of a loyalty card.

How much did she have to pay per tube of toothpaste? Express your answer as a percentage of the original price.

A. 16.67%   B. 33%   C. 60%   D. 66.7%   E. 100%

Question 18:

"You can remain fit throughout life if you exercise regularly. Simon does not exercise regularly, so he can never become fit."

Which flawed argument has the same structure as this?

A. "You can speak a foreign language if you learn when young. Simon does not speak a foreign language, so he did not learn when young."

B. "You are never tired if you sleep for 8 hours a night. Simon is tired, therefore he doesn't sleep for 8 hours a night"

C. "You can be a good musician if you practice regularly. Simon does not practice regularly, so he can never be a good musician."

D. "You can be good at sport if you have a natural ability. Simon is good at hockey, therefore he has a natural ability."

E. "Eating five portions of fruit and vegetables daily reduces the risk of heart disease. Simon eats more than this, so he will not develop heart disease."

Question 19:

"Reports of cybercrime are increasing year on year. Last year, police dealt with 250% more cybercrime then the year before. Common complaints relate to inappropriate or defamatory use of social media. To deal with this, many police forces are creating dedicated teams to deal with online offences. A pilot study showed that a dedicated cybercrime team solved cases of cybercrime 40% faster than regular detectives. Therefore the measure will act to suppress the rise in cybercrime."

Which statement best validates the above argument?

A. Solving crimes faster is necessary to keep pace with the increase in crime

B. Solving crimes faster leads to more convictions

C. Solving crimes faster increases police resources to tackle crime

D. Solving crimes faster saves money

E. Solving crimes faster reassures the public of action

Question 20:

"Recently in Kansas, a number of farm animals have been found killed in the fields. The nature of the injuries is mysterious, but consistent with tales of alien activity. Local people talk of a number of UFO sightings, and claim extra terrestrial responsibility. Official investigations into these claims have dismissed them, offering rational explanations for the reported phenomena. However, these official investigations have failed to deal with the point that, even if the UFO sightings can be explained in rational terms, the injuries on the carcasses of the farm animals cannot be. Extra terrestrial beings must therefore be responsible for these attacks."

Which of the following best expresses the main conclusion of this argument?
A. Sightings of UFOs cannot be explained by rational means
B. Recent attacks must have been carried out by extraterrestrial beings
C. The injuries on the carcasses are not due to normal predators
D. UFO sightings are common in Kansas
E. Official investigations were a cover-up

Question 21:

"To make a cake you must prepare the ingredients and then bake it in the oven. You purchase the required ingredients from the shop, however the oven is broken. Therefore you cannot make a cake."

Which of the following arguments has the same structure?
A. To get a good job, you must have a strong CV then impress the recruiter at interview. Your CV was not as good as other applicants, therefore you didn't get the job.
B. To get to Paris, you must either fly or take the Eurostar. There are flight delays due to dense fog, therefore you must take the Eurostar.
C. To borrow a library book, you must go to the library and show your library card. At the library, you realise you have forgotten your library card. Therefore you cannot borrow a book.
D. To clean a bedroom window, you need a ladder and a hosepipe. Since you don't have the right equipment, you cannot clean the window.
E. Bears eat both fruit and fish. The river is frozen, so the bear cannot eat fish.

Question 22:

"Growing vegetables requires patience, skill and experience. Patience and skill without experience is common – but often such people give up prematurely as skill alone is insufficient to grow vegetables, and patience can quickly be exhausted."

Which of the following summarises the main argument?
A. Most people lack the skill needed to grow vegetables
B. Growing vegetables requires experience
C. The most important thing is to get experience
D. Most people grow vegetables for a short time but give up due to a lack of skill
E. Successful vegetable growers need to have several positive traits

Question 23:

"Joseph has a bag of building blocks of various shapes and colours. Some of the cubic ones are black. Some of the black ones are pyramid shaped. All blue ones are cylindrical. There is a green one of each shape. There are some pink shapes."

*Which of the following is definitely **NOT** true?*
A. Joseph has pink cylindrical blocks
B. Joseph doesn't have pink cylindrical blocks
C. Joseph has blue cubic blocks
D. Joseph has a green pyramid
E. Joseph doesn't have a black sphere

Question 24:

Sam notes that the time on a normal analogue clock is 1540hrs. What is the smaller angle between the hands on the clock?

A. 110°   B. 120°   C. 130°   D. 140°   E. 150°

Question 25:

A fair 6-faced die has 2 sides painted red. The die is rolled 3 times. What is the probability that at least one red side has been rolled?

A. $^8/_{27}$ B. $^{19}/_{27}$ C. $^{21}/_{27}$ D. $^{24}/_{27}$ E. 1

Question 26:

"In a particular furniture warehouse, all chairs have four legs. No tables have five legs, nor do any have three. Beds have not less than four legs, but one bed has eight as they must have a multiple of four legs. Sofas have four or six legs. Wardrobes have an even number of legs, and sideboards have an odd number. No other furniture has legs. Brian picks a piece of furniture out, and it has six legs."

What can be deduced about this piece of furniture?

A. It is a table
B. It could be either a wardrobe or a sideboard
C. It must be either a table or a sofa

D. It must be either a table, a sofa or a wardrobe
E. It could be either a bed, a table or a sofa

Question 27:

Two friends live 42 miles away from each other. They walk at 3mph towards each other. One of them has a pet pigeon which starts to fly at 18mph as soon as the friends set off. The pigeon flies back and forth between the two friends until the friends meet.

How many miles does the pigeon travel in total?

A. 63   B. 84   C. 114   D. 126   E. 252

Question 28:

"Fruit juice contains fibre, vitamins and minerals and can be part of a healthy diet. However, it has been suggested that the high sugar content and acidity negates these benefits by leading to increased rates of dental cavities and hyperactivity in children. If left unchecked, a combination of poor dental hygiene and inappropriate diet can lead to disastrous consequences, including serious infections. On the other hand, many juices contain essential vitamins such as vitamin C which helps the immune system fight infections."

What is the main message from this passage?

A. Children should not drink fruit juice.
B. Fruit juice is harmful to health.
C. Fruit juice is good for health.

D. On balance, we should drink more fruit juice.
E. The overall benefits of fruit juice are unclear.

Question 29:

A complete stationery set includes a pen, a pencil, a geometry set and a pad of paper. Pens cost £1.50, pencils cost 50p, geometry sets cost £3 and paper pads cost £1. Sam, Dave and George each want complete sets, but Mr Browett persuades them to share. Sam and Dave agree to share a paper pad and a geometry set. George must have his own pen, but agrees that he and Sam can share a pencil.

What is the total amount spent?

A. £12.00   B. £13.50   C. £16.50   D. £17.50   E. £18.00

Question 30:

"If the government financially supports the arts, a proportion of each person's taxes will be used to finance museums, galleries and theatres. But some taxpayers have no interest in the arts and never go to theatres or museums. Many of those who enjoy the arts are able to afford to pay for them. Since no one should be forced to subsidise services which they themselves do not use, taxpayers' money should not be used to support the arts."

Which counter-argument provides the strongest rebuke of this principle?

A. If public funding for the arts is withdrawn, only those who are genuinely interested would pay to visit museums

B. The rail network is publically subsidised, although some people do not use trains

C. If people only pay for services they use, then those who can afford private health insurance would not pay towards the NHS

D. Funding museums allows greater preservation of our heritage

E. If something requires subsidy, then people must not genuinely want it

Question 31:

The figure below shows 5 squares made from 12 matches. Which 2 matches need to be moved to make 7 squares?

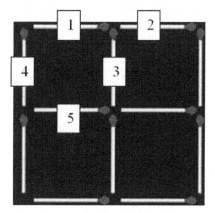

A. 1 and 2

B. 1 and 3

C. 1 and 4

D. 1 and 5

E. Not possible

Question 32:

A cube has six sides of different colours. The red side is opposite to black. The blue side is adjacent to white. The brown side is adjacent to blue. The final side is yellow.

Which colour is opposite brown?

A. Red   B. Black   C. Blue   D. White   E. Yellow

Question 33:

The UK imports 36,000,000 kg of cocoa beans each year. Each g costs the UK 0.3p, from which the supplier takes 20% commission. Of what is left, the local government takes 60% and the distribution company gets 30%.

How much are the cocoa farmers left with per year?

A. £3.68m   B. £6.82m   C. £8.64m   D. £10.8m   E. £11.4m

Questions **34** and **35** refer to the following passage:

- In the year ending June 2013 there were 1,730 fatalities in reported personal injury accidents, a 3% drop from 1,785 in the year ending June 2012. The number of killed or seriously injured (KSI) casualties fell by 5%, to 23,530, and the total number of casualties fell by 7% to 188,540.

- A total of 8,560 car users were reported killed or seriously injured in the year ending June 2013, a fall of 6% from the previous 12-month period.

- KSI casualties for the vulnerable road user groups – pedestrians, pedal cyclists and motorcyclists – showed overall decreases of 7%, 1% and 6% respectively compared with the year ending June 2012.

- The casualty rate per billion vehicle miles decreased for all casualty severities in the year ending June 2013, with falls of 3% for fatalities, 6% for serious injuries and 7% for all casualties. This is the first publication in which the Department has included quarterly casualty rates.

- There were also significant decreases in the number of child casualties (aged 0-15) which fell from 18,166 in the year ending June 2012, to 15,920 in the year ending June 2013, a fall of 12%. The number of child KSIs also fell in the same period by 11% to 2,080. The number of child pedestrian casualties who were killed or seriously injured fell by 8% to 1,440 in the year ending June 2013.

- There were drops in the number of accidents on all road types in the year ending June 2013 relative to the year ending June 2012. The number of fatal or serious accidents fell by 7% on major roads (motorways and A roads) and 4% on minor roads. On roads with speed limits over 40 mph (non-built up) fatal and serious accidents fell by 6% and on roads with speeds limits up to an including 40 mph (built-up) they fell by 5%.

- There were 185,540 casualties from 139,350 accidents in the year ending June 2013 which represents a 6% fall for accidents and a 7% fall for casualties compared with the year ending June 2012.

Question 34:

Regarding the passage, which of these statements can be known to be true?
A. Child casualties are on the rise
B. Annual road deaths in the UK are falling
C. Vulnerable road users are more likely to be injured per vehicle mile than drivers
D. From June 2012 to June 2013, there were 188,540 serious injuries
E. Motorways are safer than built-up roads

Question 35:

"The government is always under pressure to reduce road casualties. For this reason, anti-drink-drive campaigns costing millions of pounds are commonly produced, particularly around Christmas time. To address a one-year increase in drink driving related deaths, a new campaign was introduced. Subsequently, drink-driving casualties fell. The government therefore concluded that the £8m campaign had been a success."

Which of the following most undermines this argument?
A. Fewer people drink-drive these days than 10 years ago
B. Correlation does not imply causation: there is no plausible mechanism for the campaign to provide benefit.
C. The effect is too rapid for this campaign to have changed the public's attitude
D. Regression to the mean can explain this phenomenon: values which were abnormally high one year are likely to settle down the next
E. When spending so much money, benefits are certain. The true test is in running a smaller campaign.

Question 36:
"Some people with a sore throat and a chest infection have the 'flu'."

Which of the following statements is supported?
A. Some people have a chest infection, but do not have the 'flu'
B. Some people with a sore throat and a chest infection do not have the 'flu'
C. Kate has the 'flu'. Therefore she has a sore throat
D. The 'flu' is defined as a sore throat and chest infection together
E. None of the above

Question 37:
Catherine has 6 pairs of red socks, 6 pairs of blue socks and 6 pairs of grey socks in her drawer. Unfortunately, they are not paired together. The light in her room is broken so she cannot see what colour the socks are. She decides to keep taking socks from the drawer until she has a matching pair. What is the minimum number of socks she needs to take from the drawer to guarantee at least one matching pair can be made?
A. 2   B. 3   C. 4   D. 5   E. 6

Question 38:
Luca and Giovanni are waiters. One month, Luca worked 100 hours at normal pay and 20 hours at overtime pay. Giovanni worked 80 hours at normal pay and 60 hours at overtime pay. Neither received any tips. Luca earned €2000; Giovanni earned €2700.

What is the overtime rate of pay?
A. €10 per hour   D. €25 per hour
B. €15 per hour   E. €30 per hour
C. €20 per hour

Question 39:
"Train A leaves Plymouth at 10:00 and travels at 90mph. Train B leaves Manchester at 10:45 and travels at 70mph. The distance between the two cities is 405 miles. Due to a mistake, both trains are travelling on the same track."

Calculate the distance from Plymouth at which the trains will collide.
A. 158 miles   B. 203 miles   C. 228 miles   D. 248 miles   E. 263 miles

Question 40:
"100 pieces of rabbit food will feed one pregnant rabbit and two normal rabbits for a day. 175 pieces of food will feed two pregnant and three normal rabbits for a day. There is no excess food."

Which statement is NOT true?
A. A normal rabbit can be fed for longer than a day with 30 pieces of food.
B. 70 pieces of food are sufficient to feed a pregnant rabbit for a day.
C. A pregnant rabbit needs twice as many pieces per day as a normal rabbit.
D. Two pregnant and four normal rabbits will need 200 pieces of food for a day.
E. Three pregnant and ten normal rabbits will need 450 pieces of food for a day.

Question 41:
 "Studies of the brains of London taxi drivers show that training for "the knowledge", a difficult exam requiring knowledge of 20,000 London streets, enlarged a part of the brain believed to be important for spatial and organisational memory. This shows the brain can adapt to training and increase its abilities. Therefore if I wanted to improve my ability to remember names, I should also train my brain with repetitive tasks."

*Which of the following **best** represents the flaw in this argument?*
A. Enlarging of the brain does not necessarily mean it has improved
B. It might not be true to assume name memory and spatial memory use the same part of the brain
C. The brain enlargement would likely have happened anyway even without training
D. We do not know how London taxi drivers prepare for "the knowledge"
E. Practice does not necessarily improve performance on memory tasks

Question 42:
 "Michael bought a painting at an auction for £60. After 6 months, he realised the value of the painting had increased, so he sold it for £90. Realising a mistake, he wanted to buy the painting back, which he was able to do for £110. A year later, he then re-sold the painting for £130."

What is the total profit on the painting?
A. £20   B. £30   C. £40   D. £50   E. £60

Question 43:
 "Insect pests such as aphids and weevils can be a problem for farmers, as they feed on crops, causing destruction. Thus many farmers spray their crops with pesticides to kill these insects, increasing their crop yield. However, there are also predatory insects such as wasps and beetles that naturally prey on these pests – which are also killed by pesticides. Therefore it would be better to let these natural predators control the pests, rather than by spraying needless chemicals."

Which of the following best describes the flaw in this logic?
A. Many pesticides are expensive, so should not be used unless necessary
B. It fails to consider other problems the pesticides may cause
C. It does not explain why weevils are a problem
D. It fails to assess the effectiveness of natural predators compared to pesticides
E. It does not consider the benefits of using fewer pesticides

Question 44:
A parliament contains 400 members. Last election, there was a majority of 43% of the popular vote to the liberal party. However, as a first-past-the-post system of constituencies was in effect, they gained 298 seats in parliament.

How many excess members did they have, relative to a straight proportional representation system?
A. 72   B. 98   C. 112   D. 126   E. 148

Question 45:
A cube is painted such that no two faces that touch may be the same colour. What is the minimum number of colours required for this?
A. 2   B. 3   C. 4   D. 5   E. 6

Question 46:

4 people need to cross a river, and one of them is fat. They make a stable raft, but find it can only take the weight of either two thin people or the fat person alone. The raft must have someone in it to cross the river in order to propel and steer it.

What is the minimum number of journeys the raft must make across the river to get all 4 people to the other side?

A. 3   B. 5   C. 7   D. 9   E. 11

Question 47:

In a given year, there were four Wednesdays and four Saturdays in December. What day was Christmas Day (25th December)?

A. Monday   D. Thursday
B. Tuesday   E. Friday
C. Wednesday

Question 48:

"Ruddock is West of Langley but East of Dell. Hampton is midway between Langley and Ruddock. Iver is West of Ruddock. Johnstown is not East of Langley."

Which of the following cannot be concluded?

A. Hampton is East of Iver and Ruddock.
B. Ruddock is West of Langley and East of Iver.
C. Dell is west of Hampton and Langley.
D. Langley is East of Ruddock and East of Hampton
E. Iver is West of Ruddock and West of Dell.

Question 49:

"Zips and buttons are on the opposite side of women's clothing relative to men's. This is because high society always dictated clothing style, and women in high society would historically have had someone to dress them. Therefore the fastenings were positioned for the convenience of the servant and not the wearer. In our age, very few people have a servant to dress them. Therefore buttons and zips on women's clothing should be moved in accordance with the style of men's clothing."

Which of the following statements best describes the principle supporting this argument?

A. The needs of the majority should be of foremost importance
B. It would be more cost effective to make all clothes the same way
C. Traditions are of little value as times change
D. It would be easier for women to fasten clothes if buttons were reversed
E. Style is no longer dictated by high society

Question 50:

"Mark does not drink tea after 9pm as it contains caffeine. Coffee contains more caffeine than tea, therefore Mark does not drink coffee after 9pm either."

Which argument has the same structure?

A. Mark does not like onions. Curry contains onions, therefore Mark does not like curry either.
B. Mark cannot afford a pure wool suit. Pure silk suits are more expensive, so Mark cannot afford a silk suit either.
C. Mark is travelling to London. Brighton is further then London, so Mark is not travelling to Brighton.
D. The bus into town is slow. Therefore Mark will take a taxi there instead.
E. Mark can run faster than Steve. Joe is not as fast as Steve, therefore Mark can run faster than Joe.

END OF SECTION

YOU MUST ANSWER <u>ONLY</u> <u>ONE</u> OF THE FOLLOWING QUESTIONS

Question 1:

"Strive not to be a success, but to be of value"

To what extent is it possible to be "a success", but to have little value?

Question 2:

Is the media a positive or negative influence on scientific understanding?

Question 3:

"Why tell the truth if a lie is better for all concerned?"

In what circumstances can dishonesty be justified?

Question 4:

"Science is a nothing more than a thought process"

What actually is science and how is it of value to us?

END OF TEST

Mock Paper B

Question 1:

"Joseph changes jobs and gets a basic pay cut of 5%, but his tax-free monthly bonus increases from £40 to £90. He also changes tax bracket, so instead of his 10% flat rate, he pays 20% tax on all income over £10,000 pa. Joseph's current weekly take-home pay is £560 per week, exclusive of bonus."

What will his new annual take-home pay be, to the nearest hundred pounds?

A. £25,800   B. £26,500   C. £27,700   D. £29,300   E. £31,300

Question 2:

"Peter books a return flight to Dubai for £725. The flight is refundable, but there is a fee of £45 payable for cancelling. Peter notices as time passes, the remaining tickets on the same plane are becoming cheaper. He decides to cancel his flight, booking a new one for £530 through the same provider. Once again he sees prices have fallen, so he cancels this flight but can only buy a new one for £495."

What is his overall saving, relative to the original price paid?

A. £110   B. £140   C. £150   D. £195   E. £230

Question 3:

"You have three bags, each containing four balls numbered with single digit numbers. Bag A contains even numbers only, Bag B contains odd numbers only, and Bag C contains the numbers 2, 5, 6 and 8. You take a ball from Bag B and put it into Bag C; then you then take a ball from Bag C and put it into Bag A. You draw a ball at random from Bag A."

What is the probability that this ball is an odd number?

A. $^1/_{25}$ B. $^2/_{25}$ C. $^3/_{25}$ D. $^4/_{25}$ E. $^1/_5$

Question 4:

The price of bread rises by 40% due to a poor grain harvest. This is later reduced by 20% due to a government farming subsidy. Dave buys three loaves of bread and gets a fourth free because of a discount in the shop. How much did he pay per loaf of bread? Express your answer as a percentage of the original price.

A. 66%   B. 84%   C. 92%   D. 98%   E. 110%

Question 5:

Sam notes that the time on a normal analogue clock is 2120hrs. What is the smaller angle between the hands on the clock?

A. 130°   B. 140°   C. 150°   D. 160°   E. 170°

Question 6:

Sam needs to measure out exactly 4 litres of water into a tank. He has two pieces of equipment – a bucket that holds 5 litres and a one that holds 3 litres, with no intermediate markings. Is it possible to measure out 4 litres?

If so, how much water is needed in total in order to measure the 4 litres?

A. 4 litres   D. 10 litres
B. 7 litres   E. Not possible with this equipment
C. 8 litres

Question 7:

"A librarian is sorting books into their correct locations. All history books belong to the right of all science books. Science books are divided into five locations: engineering, biology, chemistry, physics and mathematics (in order from right to left). Art books are located between engineering and sport, and sport books between art and history. Literature books are to the right of art books."

What can be said with certainty about the location of literature books?
A. They are located between art and history books
B. They are located to the left of history books
C. They are located between mathematics and art
D. They are located to the right of engineering
E. They are not located to the left of sport

Question 8:

"Many people choose not to buy brand new cars, as buying brand new has significant disadvantages. Most importantly, a car's value drops substantially at the moment it is first driven on the road. Even though a car is virtually unchanged by these first few miles, the potential resale value is significantly reduced. Therefore it is better to buy second hand cars, as their value does not drop so much immediately after purchase."

Which of the following best represents the main conclusion of this passage?
A. There are many equal reasons to avoid buying brand new cars
B. Cars that have driven lots of miles should be avoided
C. The rapid loss of value in new cars makes buying second-hand a wise choice
D. Second hand cars are at least as good as new ones
E. New cars should not be driven to ensure they keep their resale value

Question 9:

James is a wine dealer specialising in French wine. From his original stock of 2,000 bottles in one cellar, he sells 10% to one customer and 20% of the remaining wine to another customer. He makes £11,200 profit from the two transactions combined.

What is the average profit per bottle?
A. £18   B. £20   C. £22   D. £24   E. £26

Question 10:

"Why should we bother exploring deep into the oceans? The programmes are very expensive, and seldom produce any results which benefit normal people. Instead, we should invest resources into supporting people in trouble, rather than wasting money on needless exploration."

Which of the following, if true, would most weaken the above argument?
A. Ocean exploration is less expensive than space exploration, which people are generally happy with
B. Ocean exploration provides fascinating information about bizarre life forms
C. Exploration has led to the discovery of new chemicals which have been used for many new medically useful drugs
D. Exploring into the oceans is safe, given modern submarine technology
E. Money is useful to help people in trouble

Question 11:

"Many good quality pieces of old furniture are considered 'timeless' – they are used and enjoyed by many people today, and this is expected to continue for many generations to come. However, most of this furniture dates back to previous eras, and modern furniture does not fall under the 'timeless' category of being enjoyed for many years to come."

Which of the following is the main flaw in the argument?
A. There may be many factors which make furniture good
B. There used to be more furniture makers than today
C. No evidence is given to tell us old furniture is better than new
D. Old furniture is desirable for other reasons than its quality
E. We cannot yet tell whether new furniture will become 'timeless'

Question 12:

"Red wine is thought to be much healthier than beer because it contains many antioxidants, which have been shown to be beneficial to health. Many red wines are produced in Southern France and Italy, therefore it is no surprise that residents there have a greater life expectancy than in the UK and Germany, which are predominantly beer producing countries."

Which of the following is an assumption of the above argument?
A. Italian people drink red wine
B. Antioxidants are beneficial for health
C. British people prefer beer to red wine
D. Beer is not produced in Italy
E. Italian life expectancy is greater than in the UK

Question 13:

Hannah, Jane and Tom are travelling to London to see a musical. Hannah catches the train at 1430. Jane leaves at the same time as Hannah, but catches a bus which takes 40% longer then Hannah's train. Tom also takes a train, and the journey time is 10 minutes less then Hannah's journey, but he leaves 45 minutes after Jane leaves. He arrives in London at 1620.

At what time will Jane arrive in London?
A. 1545   B. 1600   C. 1615   D. 1700   E. 1715

Question 14:

At a show, there are two different ticket prices for different seats. The cost is £10 for a standard seat, and £16 for a premium view seat. The total revenue from a show is £6,600, and the total attendance was 600.

How many premium view seats were purchased?
A. 60   B. 100   C. 140   D. 180   E. 240

Question 15:

The moon orbits the Earth once every 28 days. Between 20th January and 23rd April inclusive, how many degrees has the Moon turned through? This is not a leap year.
A. 1010   B. 1100   C. 1210   D. 1500   E. 1620

Question 16:

Drama academies are special schools that students can go to in order to learn performing arts. These schools are only available to the most skilled young performers, and aim to give students the best training in the arts, whilst still covering mainstream academic subjects. However, many parents are reluctant for their children to attend such academies, as they feel the academic teaching will be worse than a standard school.

Which of the following, if true, would most weaken the above argument?

A. Most top actors attended a drama academy as children
B. There is as much time dedicated to academic work in drama academies as there is in normal schools
C. The academic work comprises a greater proportion of the study time than drama related activities
D. Most children are keen to attend a drama academy if given the opportunity
E. 80% of students at drama academies attain higher than average GCSE scores

Question 17:

Anil and Suresh both leave point A at the same time. Anil travels 5km East then 10km North. Anil then travels a further 1km North before heading 3km West. Suresh travels East for 2km less than Anil's total journey distance. He then heads 13km North, before pausing and travelling back 2km South. How far, as the crow flies, are the two men now apart?

A. 11km   B. 12km   C. 13km   D. 15km   E. 17km

Question 18:

Building foundations are covered by 14cm of concrete. A builder thinks this is too thick, and grinds down the concrete by an amount three times the thickness of the concrete which he eventually leaves.

What is the remaining thickness of concrete?

A. 1.5cm   B. 2.0cm   C. 2.5cm   D. 3.0cm   E. 3.5cm

Question 19:

Chris leaves his house to go and visit Laura, who lives 3 miles away. He leaves at 1730 and walks at 4mph towards Laura's house, stopping once for a 5-minute chat to a friend. Meanwhile Sarah also wants to visit Laura. She sets off from her house 6 miles away at 1810, driving in her car and averaging a speed of 24mph.

Who reaches the house first and with how long do they wait for the other person?

A. Chris, and waits 5 mins for Sarah
B. Chris, and waits 10 mins for Sarah
C. Sarah, and waits 5 mins for Chris
D. Sarah, and waits 10 mins for Chris
E. They both arrive at the same time

Question 20:

"Illegal film and music downloads have increased greatly in recent years. This causes significant harm to the relevant industries. Many people justify this to themselves by telling themselves that they are only diverting money away from wealthy and successful singers and actors, who do not need any more money anyway. But in reality, illegal downloads are deeply harming the music industry, making many studio workers redundant and making it difficult for less famous performers to make a living."

Which of the following best summarises the conclusion of this argument?

A. Unemployment is a problem in the music industry
B. Taking profits away from successful musicians does more harm than good
C. Studio workers are most affected by illegal downloads
D. Illegal downloads cause more harm than people often think
E. Buying music legally helps keep the music industry productive

Question 21:

"40,000 litres of water will extinguish two typical house fires. 70,000 litres of water will extinguish two house fires and three garden fires. There is no surplus water"

*Which statement is **NOT** true?*
A. A garden fire can be extinguished with 12,000 litres, with water to spare.
B. 20,000 litres is sufficient to extinguish a normal house fire.
C. A garden fire requires only half as much water to extinguish as a house fire.
D. Two house and four garden fires will need 80,000 litres to extinguish.
E. Three house and ten garden fires will need 140,000 litres to extinguish.

Question 22:

A car travels at $20ms^{-1}$ for 30 seconds. It then accelerates at a constant rate of $2ms^{-2}$ for 5 seconds, then proceeds at the new speed for 20 seconds before braking with constant deceleration of $3ms^{-2}$ to a stop.

What distance is covered in total?
A. 1325m   B. 1350m   C. 1375m   D. 1425m   E. 1475m

Question 23:

"Plans are in place to install antennas underground, so that users of underground trains will be able to pick up mobile reception. There are, as usual, winners and losers from this policy. Supporters of the policy argue that it will lead to an increase in workforce productivity and will increase convenience in day-to-day life. Critics respond by saying that it will lead to an annoying environment whilst travelling, it will more easily facilitate a terrorist threat and it will decrease levels of sociability. The latter camp seems to have the greater support and so a re-consideration of the policy is urged."

*Which of the following **best** summarises the conclusion of this passage?*
A. The disadvantages of installing underground antennas outweigh the benefits
B. The cost of the scheme is likely to be prohibitive
C. The policy must be dropped, since the majority do not want it
D. More people don't want this scheme than do want it
E. A detailed consultation process should take place

Question 24:

"Ecosystems in the oceans are changing. Recently, restrictions on fishing have been imposed to tackle the decline in fish populations. As a result, farm fishing and the price of fish have increased whilst the seas recover. It is hoped that these changes will lead to a brighter future for all."

*Which of the following are **TWO** assumptions of this argument?*
*PLEASE MARK **TWO** RESPONSES*
A. People will still buy farmed fish at a higher price
B. The population of wild fish can recover
C. Fishermen will benefit from working on this scheme
D. Ecosystems have been altered as a result of climate change
E. Heavy sea fishing is to blame for the changes in the ecosystem

Question 25:

Brian is tossing a coin. He tosses the coin 5 times. What is the probability of tossing exactly 2 heads?
A. $^1/_{16}$ B. $^5/_{32}$ C. $^4/_{16}$ D. $^5/_{16}$ E. $^7/_{16}$

Question 26:
The amount of a cleaning powder to be added to a bucket of water is determined by the volume of water, such that exactly 40g is added to each litre. A bucket contains 5 litres of water, and is required to have cleaning powder added. However, the markings on the bucket are only accurate to the nearest 2%. Calculate the difference between the maximum and minimum amounts of cleaning powder which might be required to be added to make up the solution correctly.
A. 4g   B. 6g   C. 8g   D. 12g   E. 20g

Question 27:
International telephone calls are charged at a rate per minute. For a call between two European countries, the rate is 22p per minute off-peak and 32p per minute at peak hours, rounded up to the nearest whole minute. In addition, there is a connection fee of 18p for every call.

What is the cost of an off-peak call from France to Germany, lasting 1.4 hours?
A. £18.48   B. £18.66   C. £26.88   D. £27.06   E. £30.98

Question 28:
"UV radiation is harmful to the skin, and can lead to the development of skin cancers. Despite this, many people sunbathe and use tanning salons, exposing themselves to dangerous radiation. If people took more sensible decisions about their health, many serious diseases, such as skin cancers, could be avoided."

What is the main conclusion of this passage?
A. UV radiation is harmful to the skin
B. Many people like to get tanned, despite the risks
C. People do not always consider the health risks of choices they make
D. Skin cancer is a serious disease
E. Sunbathing is risky, and people should avoid it

Question 29:
"Today it is raining. My umbrella is broken; therefore I will wear an anorak."

Which of the following arguments follows the same structure?
A. I want to go to London. The train is delayed, therefore I will be late
B. The end-of-year exam is difficult. I want to do well, so will study hard.
C. My clothes are wet. The tumble dryer is in use, therefore I will pin the clothes on the washing line.
D. The piano is old. Because it is out of tune, I will not play it.
E. Bleach is an effective cleaning product. However there is also soap, which is better for washing hands.

Question 30:
Jim washes windows for pocket money. Washing a window takes two minutes. Between one house and the next, it takes Jim 15 minutes to pack up, walk to the next house and get ready to start washing again. Each resident pays Jim £3 per house, regardless of how many windows the house has. In one day, Jim washes 8 houses, with an average of 11 windows per house.

What is his equivalent hourly pay rate?
A. £4.38   B. £4.86   C. £5.12   D. £5.62   E. £6.12

Question 31:

"Bottled water is becomingly increasingly popular, but it is hard to see why. Bottled water costs many hundreds of times more than a virtually identical product from the tap, and bears a significant environmental cost of transportation. Those who argue in favour of bottled water may point out that the flavour is slightly better – but would you pay 300 times the price for a car with just a few added features?"

Which of the following, if true, would most weaken the above argument?
A. Bottled water has many health benefits in addition to tasting nicer
B. Bottled water does not taste any different to tap water
C. The cost of transportation is only a fraction of the costs associated with bottling and selling water
D. Some people do buy very expensive cars
E. Buying bottled water supports a big industry, providing many jobs to people

Question 32:

"There are no marathon runners that aren't skinny, nor no cyclists that aren't marathon runners."

*Which of the following **must** be true?*
A. Cyclists do not run marathons
B. Cyclists are all skinny
C. Any skinny person is also a cyclist
D. Marathon runners must all be cyclists
E. All of the above

Question 33:

"Langham is East of Hadleigh but West of Frampton. Oakton is midway between Langham and Stour. Frampton is West of Stour. Manley is not East of Langham."

*Which of the following **cannot** be concluded?*
A. Oakton is East of Langham and Hadleigh.
B. Frampton is West of Stour and East of Manley.
C. Stour is East of Hadleigh and Langham.
D. Oakton is East of Langham and West of Frampton
E. Manley is West of Oakton and West of Frampton.

Question 34:

A pot of paint gives sufficient paint to cover 12m^2 of wall area. The inner surface of a planetarium must be painted. The planetarium consists of a hemispheric dome of internal diameter 14 metres. How many pots of paint are required to give the dome two full coats of paint? [Assume π=3]; [Surface area of sphere = $4\pi r^2$].
A. 25   B. 36   C. 49   D. 64   E. 98

Question 35:

A planetarium has just been painted as in **34**, above. Assuming each pot of paint is 2 litres, and that the solid component of the paint is 40%, calculate the percentage decrease in the volume of the planetarium, due to the painting.
[Assume π=3]; [Volume of sphere = $4/3\pi r^3$]; [1 litre = 0.001m^3].
A. 0.0029%   B. 0.0057%   C. 0.029%   D. 0.057%   E. 2.86%

Question 36:

A sweet wrapping machine takes 400ms to wrap a sweet. How many sweets can it wrap in 2 hours?
A. 3,000   B. 7,000   C. 9,000   D. 14,000   E. 18,000

Question 37:

John's brother is 6 years younger than him. In 8 years time, the sum of their ages will be 52. How old is John now?

A. 15   B. 18   C. 21   D. 24   E. 26

Question 38:

"In the UK there is currently a housing shortage. This has increased the prices of many homes, making it harder for first time buyers to get on the property ladder than ever before. Some people attribute this to the increased divorce rate and the breakdown of the family unit as with more people living on their own, the number of occupied houses is increased. To solve the issue, more new housing needs to be built."

Which of the following best expresses the main conclusion of this passage?

A. There are not enough houses in the UK
B. Building more new housing will reduce house prices
C. The increased divorce rate is the main reason for housing shortages
D. The current rate of house building is low
E. Divorcees often live in the type of house first time buyers desire

Question 39:

A train travels from Crabtree to Eppingsworth. There are four stations in between at which the train stops. The time taken to travel between each of these stations decreases by one-fifth for each leg of the journey. Travelling from Station 3 to Station 4 takes 16 minutes.

What is the total journey time?

A. 80 minutes   D. 125 minutes
B. 89 minutes   E. 183 minutes
C. 105 minutes

Question 40:

HS2 is a proposal for a new high speed rail link between London and Birmingham. Supporters say it is required to increase capacity on the congested existing line, and will bring economic benefits through decreased journey times. However many of those living near the proposed development are concerned. They fear the construction will bring a lot of noise and be a visual scar on the landscape, in the same way that the refurbishments of the old line did.

Which of the following, if true, would most appease the residents?

A. The benefits of fast rail links would be felt by all
B. If the line was moved elsewhere, other people would be similarly affected
C. Journey times will be reduced significantly
D. Modern trains are quieter than old ones
E. Since the last works, construction equipment has become quieter and there are more stringent regulations against noise and mess

Question 41:

Van hire from Tony's costs £23 per day. However longer term van hire is cheaper from Adam's, at a rate of £18 per day, but with an initial rental fee of £65.

How many days would you have to hire for to make a saving by hiring from Adam's rather than Tony's?

A. 9 days   B. 11 days   C. 13 days   D. 14 days   E. 17 days

Question 42:

"Antibiotic resistance is on the increase. As a result, many of the antibiotics in our vast armoury are becoming ineffective against common infections. Probably the most significant reason for this is the use of antibiotics in farming, as this exposes bacteria to antibiotics for no good reason, giving the opportunity for resistance to develop. If this worrying trend continues, we might, in 30 years' time, be back in the Victorian situation, where people die from skin or chest infections we consider mild today."

Which of the following best represents the overall conclusion of the passage?

A. Antibiotic resistance is a serious issue
B. Antibiotics use in farming is essential
C. The use of antibiotics in farming could cause us serious harm
D. Victorians used to die from diseases we can treat today
E. Antibiotics can treat skin infections

Question 43:

In a different language, the word for "AMBULANCE" is conveniently written so that it reads the same when viewed directly, and through the rear view mirror of a car. Which of the following is a viable translation for "AMBULANCE"?

        1) TANANAT
        2) AMATAMA
        3) MARAM
        4) ITRARTI
        5) SOOVOOS

A. 1 only   B. 1 and 3   C. 2 only   D. 2 and 5   E. 1, 2 and 5

Question 44:

Sam and Pete are vegetable merchants. Sam buys the vegetables and Pete sells them on. Their trading across 3 weeks is summarised here. In the first week, Sam buys £7,500 of vegetables and Pete sells these for £10,500. In week 2, Sam buys 60% more in value, and Pete sells these with a 30% profit margin. In week 3 the total sale value is £2,000 less than in week 2 – but the profit margin is a healthy 60%.

What value of vegetables did Sam buy in week 3?

A. £7,750   B. £8,500   C. £8,900   D. £9,200   E. £9,750

Question 45:

The following net folds to make a cuboctahedron. This shape is folded up and made into a die. The area of each triangle is $3\sqrt{27}$cm². Assuming the probability of landing on a side is directly proportional to the surface area of that side, calculate the probability of landing on side **X**, expressed in its simplest form.

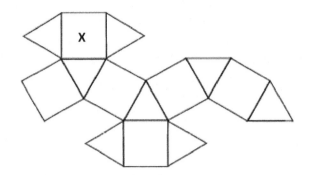

A. $\dfrac{4}{24 + \sqrt{27}}$ B. $\dfrac{36}{216 + 24\sqrt{27}}$ C. $\dfrac{9}{54 + \sqrt{27}}$ D. $\dfrac{1}{10}$ E. $\dfrac{24\sqrt{27}}{14^2}$

Question 46:

"Many people commute to work – that is to say they have a repetitive journey between their home and work. In the UK, over 50% of commuters complain about the duration of their commute. The median length of a commute in the UK is 40 minutes, which isn't that long if you think about it. Many of us waste more time than that every day for other reasons, and on the train in particular the time can be used productively."

Which of the following can be reliably concluded from this passage?
A. British people like to complain, even when there are no real problems
B. At least 50% of commuters have a commute which is not longer than 40 minutes
C. People with a commute over 40 minutes are likely to complain about it
D. 50% of people feel the distance to work is too great
E. More than half the population complain about their commute

Question 47:

In a game of bagatelle, Josie scores zero once in every three turns. She takes three turns. What is the probability she scores zero at least once?
A. $^{13}/_{27}$ B. $^{15}/_{27}$ C. $^{19}/_{27}$ D. $^{24}/_{27}$ E. 1

Question 48:

Study this graph, showing the filling of a type of container. Which type of container is being filled?

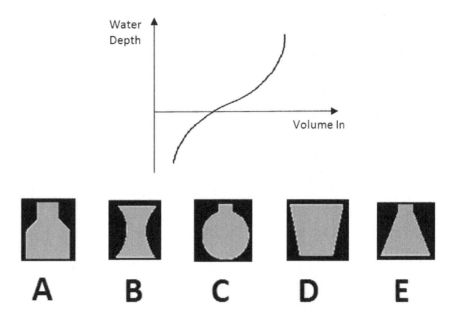

Question 49:

"There has been an increase in the number of listed buildings in the UK. Listed buildings are registered buildings of aesthetic or historical interest, which have restrictions placed on making changes to the building. Therefore it is bad to own a listed building, as it restricts the freedom with which you can modify your house as you wish.

Which of the following is the main flaw in the above argument?
A. Listed buildings are not used as homes
B. It assumes people want to modify their houses
C. It does not take account of the benefits of owning a listed building, such as funding for restorations or higher market value
D. There is no evidence that modifying a listed building is any more difficult
E. Buildings of historical interest are unlikely to need modifications

Question 50:

Scott believes the number of Ford Escorts on the roads decreases by 25% every year. In 2005, there were 36,000 hard top Escorts remaining.

How many does Scott expect will remain in 2015?
A. 1,450   B. 2,030   C. 2,420   D. 2,700   E. 3,600

END OF SECTION

YOU MUST ANSWER <u>ONLY</u> <u>ONE</u> OF THE FOLLOWING QUESTIONS

Question 1:
Design an experiment to deduce the sensitivity of a snake's hearing. Explain everything you would do, and your rationale for doing so.

Question 2:
"The eternal mystery of this world is its comprehensibility"
> To what extent is the world comprehensible?

Question 3:
"The greatest obstacle to learning is education"
> Argue for or against this statement.

Question 4:
Does a vacuum really exist?

END OF TEST

Mock Paper C

Question 1:

"Every year, there are tens of thousands of motor crashes, causing a serious number of deaths, representing the leading cause of death in the UK that is not a disease. However, in spite of this horrendous statistic, there are still thousands of uninsured drivers. The government is under a moral obligation to clamp down on uninsured drivers, to reduce the incidence of such crashes. That they have not acted is arguably the most outrageous failing of the present government."

Which of the following is the best statement of a flaw in this passage?

A. It has made unsupported claims that the government's failure to act is morally outrageous.
B. It has not provided any evidence to support its claims that motor crashes are the leading cause of death in the UK outside of diseases.
C. Even if motor crashes were prevented, it would not save lives of people who die from other causes.
D. It has implied that lack of insurance is related to the incidence of motor crashes.
E. It has fabricated an obligation on the government's part to intervene and reduce the numbers of uninsured drivers.

Question 2:

"Several years ago the Brazilian government held a referendum of the populace, to decide whether they should enact a law banning the ownership of guns. The Brazilian people voted strongly against this proposal. When asked why this had happened, one commentator said he believed the reason was that 90% of criminals who use guns to commit crimes buy their weapons on the black market, illegally. Thus, if Brazil were to ban the legal sale of guns, this would remove the ability of law-abiding citizens to purchase protection, whilst doing little to remove weapons from the hands of criminals.

Some commentators have pointed to this reasoning, and claimed that the UK should also legalise guns, to allow citizens to protect themselves. However, in the UK the black market for weapons is not as widespread as in Brazil. Thus, most people in the UK have little reason to fear gun attacks, and legalising the sale of guns would simply make it much easier for criminals to acquire weapons. Therefore, the situation in Brazil is not applicable to the UK, and legalising gun ownership in the UK would be a bad move."

Which of the following is best supported by this passage?

A. The UK should not legalise guns.
B. The UK should legalise guns.
C. Brazil should ban the ownership of guns.
D. Brazil should not ban the ownership of guns.
E. None of these statements are supported by the passage.

Question 3:

Hannah is buying tiles for her new bathroom. She wants to use the same tiles on the floor and all 4 walls, and for all the walls to be completely tiled apart from the door. The bathroom is 2.4 metres high, 2 metres wide and 2 metres long, and the door is 2 metres high, 80cm wide and at the end of one of the 4 identical walls. The tiles she wants to use are 40cm x 40cm.

How many of these tiles does she need to tile the whole bathroom?
A. 110   B. 120   C. 135   D. 145   E. 150

Question 4:

Jane and Trevor are both travelling south, from London to York. Jane is driving, whilst Trevor is travelling by train. The speed limit on the roads between York and London is 70mph, whilst the train travels at 90mph. Thus, we should expect that Trevor will arrive first.

Which of the following would weaken this passage's conclusion?

A. The train takes a more direct route, whilst the road from York to London goes through several major cities and zig-zags somewhat on its way down the country.

B. Trevor left before Jane.

C. Jane is a conscientious driver, who never exceeds the speed limit.

D. Trevor's train makes a lot of stops on the way, and spends several minutes at each stop waiting for new passengers to board.

E. Meanwhile, Raheem is making the same journey by plane, and will arrive before either Trevor or Jane.

Question 5:

A recipe for 20 cupcakes requires 200g of butter, 200g of sugar, 200g of flour and 4 eggs. Jeremy has two 250g packs of butter, a bag of 600g of sugar, a kilogram bag of flour and a pack of 12 eggs. How many cupcakes can he make and how many eggs does he have left over?

A. 50, 2   B. 50, 3   C. 60, 0   D. 60, 2   E. 60, 3

Question 6:

ABC taxis charges a rate of 15p per minute, plus £4. XYZ taxis charges a rate of £4, plus 30p per mile. I live 6 miles from the station. What would the taxi's average speed have to be on my journey home from the station for the two taxi firms to charge exactly the same fare?

A. 20mph   B. 30mph   C. 40mph   D. 50mph   E. 60mph

Question 7:

"King Arthur has been issued a challenge by Mordac, his nephew who rules the adjacent Kingdom. Mordac has challenged King Arthur to select a knight to complete a series of challenging obstacles, battling a number of dark creatures along the way, in a test known as the Adzol. The King's squire reports that there are tales told by the elders of the court meaning that only a knight with tremendous courage will succeed in Adzol, and all others will fail. He therefore suggests that Arthur should select Lancelot, the most courageous of all Arthur's Knights. The squire argues that due to what the Elders have said, Lancelot will succeed in the task, but all others will fail."

Which of the following is NOT an assumption in the squire's reasoning?

A. Lancelot has sufficient courage to succeed in the Adzol.

B. No other knights in Arthur's command also have tremendous courage, so will all fail Adzol.

C. Great courage is required to be successful in the Adzol.

D. The tales told by the elders of the court are correct.

E. None of the above – they are all assumptions.

Question 8:

A historian is examining a recently excavated hall beneath a medieval castle. She finds that there are a series of arch-shaped gaps along one length of the wall, surrounded by a different pattern of bricks to that seen elsewhere in the walls. These are found to represent where windows were once located, looking out onto one side of the castle. However, the site is now underground. Underground halls in castles never contain windows, so the historian reasons that this hall must once have been located above the ground. Therefore, the ground level must have changed since this castle was built.

Which of the following represents the main conclusion of this passage?

A. Windows are never found in underground halls.
B. Arch-shaped gaps always indicate that windows were once present.
C. It is unexpected for windows to be found in halls in castles.
D. The hall was once located above ground.
E. The ground level must have changed since this hall was built.

Question 9:

Adam's grandmother has sent him to the shop to buy bread rolls. Usually, bread rolls are 30p for a pack of 6 and so his grandmother has given him the exact amount to buy a certain number of bread rolls. However, today there is a special offer whereby if you buy 3 or more packs of rolls, the price per roll is reduced by 1p. He can now buy 1 more pack than before and get no change.
How many bread rolls was he originally supposed to buying?

A. 4   B. 5   C. 6   D. 24   E. 30

Question 10:

"The England men's cricket team have recently been knocked out of the world cup after a very poor performance that saw them eliminated at the group stage, managing only 1 win and losing against teams well below them in the rankings. The board of English cricket is sitting down to discuss why the team's performance was so poor, and what can be done to ensure that future world cups have a more positive outcome. The chairman of the board says that the current crop of players is not good enough, and that the team's performance should improve soon, as more able players come through the ranks in the county teams, so no action is needed.

However, the sporting director takes a different view, saying that England have not gone further than the group stage of any cricket world cup for the last 25 years, during which time numerous players have come and gone from the team. The sporting director argues that this long period of poor performance indicates that there is a problem with English cricket, meaning that not enough talented players are being produced in the country. He argues that therefore, steps should be taken to reform English cricket to actively foster the development of more talented players."

Which of the following, if true, would most strengthen the sporting director's argument?

A. The English cricket team is currently regarded as one of the best in the world, with some of the most talented players.
B. England have been steadily falling lower in the world cricket rankings for the last 25 years, due to poor performances across the board in various cricket competitions.
C. A skilled batsman, who was ranked as the 4th best player in the world, has recently retired from the England team. Now, there are no English cricket players in the top 10 of the world cricket player rankings, which is the first time this has happened in over 70 years.
D. Despite not performing well in world cups, England have performed well in other cricket competitions over the last 20 years.
E. Cricket was invented in England, so everybody expects that England should have a lot of good players in their team.

Question 11:

Karl is making cupcakes for a wedding. It takes him 25 minutes to prepare each batch of cakes. Only 12 can go in the oven at a time and each batch takes 20 minutes in the oven. If he needs to make 100 cupcakes by 4pm, at what time should he start?

A. 11:55am   B. 12:20pm   C. 12:40pm   D. 13:20pm   E. 14:00pm

Question 12:

	Boys Absenteeism	Girls Absenteeism	Pupils on roll	Average
Hazelwood Grammar	7%	Boys' school	300	7%
Heather Park Academy	5%	6%	1000	5.60%
Holland Wood Comprehensive	5%	6%	500	5.60%
Hurlington Academy	Girls' school		200	
Average		7%		

Some of the information is missing from the above table. What is the rate of girls' absenteeism at Hurlington Academy?

A. 6.5%   B. 7%   C. 9%   D. 11.5%   E. 13%

Question 13:

"Up until the 20th century, all watches were made by hand, by watchmakers. Watchmaking is considered one of the most difficult and delicate of manufacturing skills, requiring immense patience, meticulous attention to detail and an extremely steady hand. However, due to the advent of more accurate technology, most watches are now produced by machines, and only a minority are made by hand, for specialist collectors. Thus, some watchmakers now work for the watch industry, and only perform *repairs* on watches that are initially produced by machines. "

Which of the following <u>cannot</u> be reliably concluded from this passage?

A. Most watches are now produced by machines, not by hand.
B. Watchmaking is considered one of the most difficult of manufacturing skills
C. Most watchmakers now work for the watch industry, performing repairs on watches rather than producing new ones.
D. The advent of more accurate technology caused the situation today, where most watches are made by machines.
E. Some watches are now made by hand for specialist collectors.

Question 14:

"Many vegetarians claim that they do not eat meat, poultry or fish because it is unethical to kill a sentient being. Most agree that this argument is logical. However, some Pescatarians have also used this argument, that they do not eat meat because they do not believe in killing sentient beings, but they are happy to eat fish. This argument is clearly illogical. There is powerful evidence that fish fulfil just as much of the criteria for being sentient as do most commonly eaten animals, such as chicken or pigs, but that all these animals lack certain criteria for being "sentient" that humans possess. Thus, Pescatarians should either accept the killing of beings less sentient than humans, and thus be happy to eat meat and poultry, or they should not accept the killing of any partially sentient beings, and thus not be happy to eat fish."

Which of the following best illustrates the main conclusion of this passage.

A. The argument that it is unethical to eat meat due to not wishing to kill sentient beings but eating fish is acceptable, is illogical.
B. Pescatarians cannot use logic.
C. Fish are just as sentient as chicken and pigs, and all these beings are less sentient than humans.
D. It is not unethical to eat meat, poultry or fish.
E. It is unethical to eat all forms of meat, including fish and poultry.

Question 15:

"Recent research into cultural attitudes in Britain has revealed a striking hypocrisy. When asked whether foreign people travelling to Britain on holiday should learn some English, 60% of respondents answered yes. However, when asked if they would attempt to learn some of the language before travelling to a country which did not speak English, only 15% of the respondents answered yes. This is a shocking double-standard on the part of the British public, and is symptomatic of a deeper underlying issue that British people feel themselves superior to other cultures."

Which of the following can be reliably concluded from this passage?

A. 60% of people in Britain think that foreign people travelling to Britain for a holiday should learn English, but would not learn the language themselves when going on holiday to a country which did not speak English.
B. The British public do not feel that it is important to learn some of the language before travelling to a country which does not speak English.
C. There are numerous issues of racism amongst the British public, stemming from the fact they feel themselves superior to other cultures.
D. Less than 10% of the British public would attempt to learn some of the language before travelling to a country which did not speak English.
E. Some in Britain think that foreign people travelling to Britain for a holiday should learn English, but would not learn the language themselves when going on holiday to a country which did not speak English.

Question 16:

Harriet is a headmistress and she is making 400 information packs for the sixth form open evening. Each information pack needs to have 2 double sided sheets of A4 of general information about the school. She also needs to produce 50 A5 single sided sheets about each of the 30 A Level courses on offer. Single sided A5 costs £0.01 per sheet to print. Double sided costs twice as much as single sided. A4 printing costs 1.5 times as much as A5.

How much does she spend altogether on the printing?

A. £270   B. £310   C. £350   D. £390   E. £430

Question 17:

"Kirkleatham Town football club are currently leading the league. One week they play a crucial match against Redcar Rovers, who are second placed. The points tally of the teams in the table means that if Kirkleatham Town win this game, they will win the league. Before the game, the manager of Kirkleatham Town says that Redcar Rovers are a tough opponent, and that if his team do not play with desire and commitment they will not win the game. After the game, the manager is asked for comment on the game, and says he was pleased that his team played with so much desire, and showed high levels of commitment. Therefore, Kirkleatham will win the league."

Which of the following best illustrates a flaw in this passage?

A. It has assumed that Kirkleatham will not win the game if they do not play with desire and commitment.

B. It has assumed that if Kirkleatham play with desire and commitment, they will win the game.

C. It has assumed that Kirkleatham played with desire and commitment.

D. It has assumed that Redcar Rovers are a tough opponent, and that Kirkleatham will not be able to easily win the game.

E. It has assumed that if Kirkleatham win the match against Redcar Rovers, they will win the league.

Question 18:

Two councillors are considering planning proposals for a new housing estate, to be built on the edge of Bluedown Village. Councillor Johnson argues for a proposal to be built upon brownfield land, land which has previously been built on, rather than greenbelt land, which has not previously been built on. He argues that this will both lower the cost of building the estate, as the land would already have some underlying infrastructure and would not need as much preparation, and will ensure a minimal impact on wildlife around the area.

Which of the following would most weaken the councillor's argument?

A. Brownfield land is often not as appealing as greenbelt land visually, and it is likely that houses built on brownfield land will not sell for as high a price as houses built on greenbelt land.

B. An area of brownfield land on the edge of the village, originally built as an outdoor leisure complex, has since become run down, and ironically is now a haven for various types of rare newts, lizards and birds.

C. Much of the brownfield land around the edge of the village has undergone substantial underground development, with a good system of electricity cables, gas pipes and plumbing in place.

D. The village is surrounded by several greenbelt areas designated as areas of outstanding natural beauty, supporting an abundance of wildlife.

E. The village mayor, who has ultimate control over the planning proposal, agrees with councillor Johnson's argument. Thus, it is likely his recommendations will be followed

Question 19:

John is driving down the A1(M), southbound, having joined at Darlington. If he leaves the motorway at Junction 48, he will arrive at Wetherby Service station. If he goes to Wetherby service station, he will purchase a new cuddly soft toy for his son David. Thus, if John leaves the motorway at Junction 48, he will purchase a new cuddly soft toy.

Which of the following most closely follows the reasoning in this passage?

A. Lucy is travelling from London to Cambridge. She will have to travel on the M25 unless she travels by train. However, Lucy opts to travel by coach, so she will therefore travel on the M25.

B. James is a cricket player working on his batting technique. He finds that whenever he is delivered a ball that does not have any spin on it, he manages to score a run. Whenever he scores a run, his coach shows him a replay, so that he can have a second look at the technique he used. Thus, if James is delivered a ball without spin, he will be shown a replay by his coach.

C. A truck is delivering a cargo of fresh fruit to a supermarket. The driver knows he must deliver the shipment before 8am. If he takes the motorway, he will make the journey in 1 hour less time than if he takes the country lane. Thus, if he takes the motorway, he will be able to leave an hour later.

D. Cleveland police has been informed by the government that its funding is to be cut, and is discussing how best to deal with these cuts. The chief constable knows that if they do not maintain funding of their 999 response vehicles, then the public will be endangered. If this funding is to be maintained, then there will have to be cuts in police patrols. Thus, if the public are not to be endangered, there will have to be cuts in police patrols.

E. If a car's fuel injectors are not properly maintained, then it will be unable to deliver fuel to the engine. If it is unable to deliver fuel to the engine, it will not be able to run, as the engine cannot provide power without fuel. Thus, if the fuel injectors are not properly maintained, the car will not be able to run.

Question 20:

The graph below shows the voting intentions of some constituents interviewed by a polling group, prior to an upcoming election.

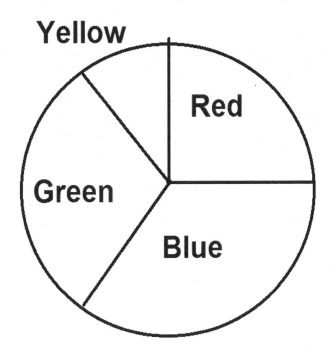

How many times more people said their intention was to vote for the red party than the yellow party?

A. 2   B. 3   C. 4   D. 5   E. 6

Question 21:

A takeaway pizza restaurant is having a sale. If you spend £30 or more at full price, you can get 40% off.
Prices are as follows:

- Basic cheese and tomato pizza: £8 small, £10 large
- All other toppings £1 each
- Sides
    - Garlic bread £3
    - Potato wedges £2.50
    - Chips £1.50
    - Dips £1 each

Ellie and Mike want to order a large pizza with mushrooms and ham, garlic bread, 2 portions of chips and a dip.

Which of these additional items can they order to minimise the price of their order?

A. Small pizza with pineapple and onion   D. 4 portions of potato wedges

B. Large pizza with mushroom   E. Garlic bread

C. Barbecue dip

Question 22:

Sohail is planning to travel between Newcastle and Middlesbrough by car next month. If Sohail does not pass his driving test before his journey, he will not be able to travel on the A1(M), because learner drivers are not allowed to travel on motorways. If he does not travel on the A1(M), he will have to travel on the A19, as this is the only other route connecting Newcastle to Middlesbrough. Thus, if Sohail does not pass his driving test, he will have to travel on the A19.

Which of the following most closely parallels the reasoning used in this passage?

A. If Anna flies from Durham Tees Valley Airport to Sweden, she will have to travel via Amsterdam, as there are no direct flights from Durham Tees Valley to Sweden. However, Anna instead chooses to fly from Manchester Airport, so she will not have to go via Amsterdam.

B. Merdoc the witch is flying on her broomstick, being pursued by a dragon who is currently flying faster than her. If she does not speed up, she will not be able to escape the dragon. If she does not escape the dragon, she will not be able to warn the high council of the impending invasion by the Mithrolites. Thus, if she does not speed up, she will not be able to warn the high council.

C. An oil company is considering purchasing a new processing plant in order to streamline their oil production process and reduce costs by 50%. If they achieve this cost reduction, they will be able to lower their petrol prices by 20%. However, the oil company does not purchase the plant, so this price reduction will not occur.

D. The Food Standards Agency (FSA) is considering implementing new labelling restrictions for genetically modified products. If they do not implement these restrictions, then companies will not have to label foods containing genetically modified produce. If labelling of genetically modified produce is not made compulsory, then environmental groups will protest, because they feel the public should be made aware what they are purchasing. Thus, if the FSA does not implement the new restrictions, environmental groups will protest.

E. In a school sports day, Rebecca is running in the final event, the 100m sprint. She knows that if she runs the sprint in less than 14 seconds, then the red group will not win gold in this event. The points have worked out such that if the red group do not win gold, then Rebecca's group will win the competition. Rebecca runs the 100m in 13.8 seconds, so her group will win the competition.

Question 23:

"The M1 Abrams tank is widely regarded as the most fearsome tank in the world. Highly advanced depleted uranium composite armour makes it difficult to damage from range, whilst a good top speed in excess of 50kmph and a large fuel capacity make it difficult to catch and contain the tank in an operational context. Whilst the tank does have weak spots that can be exploited at close range, a formidable 122m smoothbore gun as the main armament makes this an incredibly dangerous tactic for opposing tanks. Country X is developing a new main battle tank to boost the prowess of their armoured formations, and have released a statement describing how they will implement next-generation armour into this new tank, to boost its defensive capacity. The government of country X believe this will allow their new tank to compete with the best tanks in the world. However, this view is mistaken. The M1 Abrams clearly demonstrates that a *combination* of different factors, including protection, manoeuvrability and firepower, are responsible for its status as the world's most formidable tank. Simply increasing the defensive capabilities of a tank is not sufficient. Thus, Country X's government is clearly incorrect in this matter."

Which of the following best illustrates the main conclusion of this passage?
A. Increasing the defensive capacity of a tank is not sufficient to make it equal to the best tanks in the world.
B. Multiple factors are required to make a tank equal to the best tanks in the world.
C. The new tank will not be as good as the M1 Abrams, as its defensive capacity will not be as good.
D. The view of Country X's government, that increasing the defensive capacity of a tank will make it equal to the best in the world, is clearly incorrect.
E. No tank is able to compete with the M1 Abrams, which will always be the world's most formidable tank.

Question 24:

The table below shows the balances of my bank accounts in pounds. Interest is paid at the end of the calendar year. My salary, which is the same every month, is paid into my current account on the 2nd of each month. All the money I have is in one or the other of my bank accounts.

	Current Account	Savings	ISA
1st March	1300	5203	2941
1st April	3249	2948	2941
1st May	4398	9384	0
1st June	3948	8292	0

In which month did I spend the most money?
A. February
B. March
C. April
D. May
E. 2 or more months are the same

Question 25:

On Monday, my son developed a disease. No-one else in the house has the disease. The doctor gave me some medicine and told me that everyone in the house who does not have the disease should also take half the dose. We need to take the medicine for 10 days, and the dosage is based on weight.

Weight	Dosage
Under 30kg	0.1ml per kg, 3 times a day
30kg – 60kg	0.2ml per kg, 4 times a day
60kg +	0.1ml per kg, 6 times a day

My son is 40kg. I also have a daughter who is 20kg. I am 75kg and my husband is 80kg. How many 200ml bottles of medicine will we need for the whole 10 days?

A. 4   B. 5   C. 6   D. 7   E. 8

Question 26:

At Tina's nursery school, they have red, yellow and blue plastic cutlery. They have just enough forks and just enough knives for the 21 children there. There are the same number of forks as knives of each colour. Twice as many pieces of cutlery are yellow as blue. Half as many pieces of cutlery are red as blue. Tina takes a fork and a knife at random. What is the probability that she will get her favourite combination, a red fork and a yellow knife?

A. 4/49   B. 1/9   C. 36/49   D. 3/9   E. 3/49

Question 27:

"The UK's taxation and public spending is horrendously flawed, with various immoral features. One example of such a flaw is the subsidisation of public transport with money raised via taxation. According to recent research, public transport is only used by 65% of the population, and since there is no economic benefit stemming from a good public transport system, the other 45% of the population gets no benefit from public transport, but are still required to pay towards it via taxation. The system is in urgent need of reform, such that taxation is only used to support services and systems which are of benefit to everyone."

Which of the following is the best application of the principle used in this passage?

A. Only 48% of the population have ever visited an art gallery, so public funds should not be used to subsidise art galleries, as not all the population use it.

B. Primary and Secondary education provides an economic benefit to the whole country, so public funds should be used to support schools.

C. Although many people never use a hospital, we should still use public funds to provide them, because many people cannot afford private healthcare, and thus we need a publically available health service for those people.

D. There is no evidence that the fire service provides any benefit to the majority of the public, who will never experience a house fire in their lifetime. Thus, the fire service should not be publically funded via taxation.

E. The Police service is a vital service for the country so should be publically funded regardless of how few people benefit from its presence.

Question 28:

"SpicNSpan Inc is a cleaning company offering a range of cleaning services across the UK. The board has recently acquired a new chairman, who has called a meeting of the board to assess how the company can move forwards, expanding its services and increasing its market share. One of the things the new chairman is looking at is the types of services the company provides. He argues that their "All Inclusive" service, where customers pay a fixed amount to have their entire house cleaned as a one-off event, are more popular than their "Hourly" services, where customers pay for a cleaner to carry out a certain number of hours each week. The new chairman argues that they should therefore focus on the "All Inclusive services", rather than the "Hourly" services, in order to increase profits."

Which of the following best illustrates a flaw in the chairman's argument?
A. He ignores other services which may bring in even more profit than 'All Inclusive' services.
B. The fact that 'All inclusive' services are more popular than 'Hourly' services does not mean that they are more profitable. 'Hourly' services may be more profitable.
C. He has assumed that 'Hourly' services are more popular than 'All Inclusive' services.
D. He has assumed that 'All Inclusive' services are more popular than 'Hourly' services.
E. The rest of the board may have other strategies to increase profits, which are better than the new Chairman's.

Question 29:

"The effects of fossil fuels such as Oil, Coal and Natural Gas on the environment are plain and clear for everybody to see. The long-term use of such non-renewable fuels to produce power has led to devastating climate change, and will continue to cause damage as long as it continues. With this in mind, the European Commission has devised a set of targets to promote energy production by alternative types of fuels. However, there is a glaring problem with these targets. Shockingly, the Commission has targeted a "150% increase in the amount of energy produced by Nuclear Power by 2025". This is an outrageous misjudgement, because Nuclear Power is a non-renewable fuel, just like Oil, Coal and Natural Gas. If we wish to protect the environment and halt climate change, we need to switch to *renewable* fuels, which are proven not to cause damage to the environment, NOT non-renewables such as Nuclear Power."

Which of the following best illustrates a flaw in this passage?
A. It has assumed that all non-renewable power sources cause environmental damage.
B. It has assumed that renewable energy sources do not cause environmental damage.
C. It has assumed that the targets will be met, when in fact there is no guarantee that this will happen.
D. It has neglected to consider other problems with the targets set by the Commission.
E. It has assumed that the climate change caused by burning of Oil, Coal and Natural Gas cannot be offset or prevented by other strategies.

Question 30:

"Despite the overwhelming evidence which certifies that vaccines are a miracle of modern medicine, and are responsible for saving a great number of lives, there remains a stubborn section of society that refuses to be vaccinated against important diseases, insisting that they are unsafe and ineffective. This group maintain this view in spite of extremely strong evidence that vaccines are safe, and in spite of the advice given by doctors. This group is particularly strong in the USA, where they pose a very real concern. Over the last 5 years, the percentage of the population that is unvaccinated has been rising by 1 each year, such that now a staggering 6% of Americans have not received any vaccinations. Experts have advised that due to the way diseases are spread, if less than 90% of the population at any given time is unvaccinated, then it is almost certain that we will see an outbreak of Measles, a highly contagious and damaging disease. Thus, we expect that there will likely be an outbreak of measles in the next 5 years in the USA, and we should take steps to prepare for this."

Which of the following, if true, would most strengthen this argument?
A. New and powerful evidence of the safety of vaccinations is due to be released to the public next year.
B. Measles is a highly damaging disease, which frequently causes death or severe permanent injury in those affected.
C. Throughout the last half-century, the number of people who are not vaccinated has risen and fallen continuously. Usually, the increases in non-vaccinated individuals occur over a 6-year period, after which time vaccination becomes more popular, and this number falls.
D. The number of doctors advising against vaccination has been rising for the last 10 years, and shows no signs of decreasing.
E. The rise in unvaccinated individuals has been increasing steadily for 5 years. The only time such a rate of increase has occurred in history was during the 1950s/1960s. In this case, a similar rate of increase in non-vaccinated individuals was maintained for a staggering 13 years.

Question 31:

"It is well established that modern humans evolved in Africa, around 2 million years ago, and that the first humans were mainly hunter-gatherers, living off hunted meat and plant foods collected from their environment. However, this poses an interesting question. Humans are relatively weak, small, feeble creatures, and around 2 million years ago most wildlife in Africa consisted of large, powerful creatures. Thus, it is unclear how humans were able to hunt successfully, and obtain meat for food. One theory is that humans are well-built for long-distance running, largely thanks to our ability to control our temperature via sweating. This theory reasons that humans were able to pursue animals such as antelope, which run when challenged, and were able to keep on running until the antelope collapsed through heat exhaustion. Meanwhile, the humans were kept cool via sweating, and were able to then go in and butcher the defenceless antelope.

Recent evidence has emerged supporting this theory, showing that human feet are well-developed for long-distance running, with fleshy areas in the correct orientation to absorb the impact without causing joint damage, and a heart well evolved to keep pumping at a moderately fast pace for long periods. With the emergence of this powerful new evidence, we should accept this theory, known as "the persistence running theory" as true."

Which of the following identifies a flaw in this argument?
A. The emergence of evidence in support of the persistence running theory does not mean that this theory is true.
B. There is little evidence that the human body is well setup for long-distance running.
C. It has neglected to consider other theories for how humans obtained meat during their early evolution.
D. There are numerous issues with the theory of persistence running, but many of these have been resolved thanks to the new evidence that has emerged.
E. It has not considered evidence that humans evolved in Europe, where there are smaller animals which humans may have easily been able to tackle.

Question 32:

		\multicolumn{6}{c}{Predicted}					
		A	B	C	D	E	U
Actual	A	7	4	2	1	0	0
	B	3	8	2	2	1	0
	C	2	4	5	7	3	1
	D	2	2	2	6	5	0
	E	1	2	2	1	7	2
	U	1	1	0	3	5	6

The table above shows the actual and predicted AS grades of 100 students at Greentown Sixth Form. Assuming that each student can only be predicted one grade, what percentage of students had their grades correctly predicted?

A. 14%   B. 16%   C. 39%   D. 61%   E. 78%

Question 33:

In one year, Mike lowers his workers' wages by $x\%$. The next year, he lowers their wages by $x\%$. The year after this, he raises the wages by $x\%$. In the final year, he raises their wages by $x\%$. In all these stages, x is a constant positive number. Compared to the workers' original wages before any raising or lowering, what are their new wages?

A. The same as the original wages
B. Lower than the original wages
C. Higher than the original wages
D. Can't tell from the provided information even if we know what x was
E. Can't tell from the provided information but would be able to tell if we knew what x was.

Question 34:

"The medical scientific establishment has a long established system for naming body parts and medical phenomena. This system is based upon ease of understanding, such that a body part, or a process of the body, is named based on its clinical relevance. This means that features are named in a way which will help doctors understand and explain to patients what the body part is, or what is wrong with it in the case of a disease. However, this poses significant problems for scientific medical research. Often, the most important features of a body part from a scientific point of view are not the most clinically important features, leading to confusion within the scientific literature, as medical researchers misunderstand the purpose of a discussion, due to confusing nomenclature. Whilst it is important for doctors to be able to explain things clearly to patients, it is relatively easy for this to happen in spite of confusing nomenclature, whereas confusing names causes serious problems in the scientific world. Thus, the naming system for medical features should be altered, to reflect the scientifically important features of body parts, rather than the clinically important ones."

Which of the following best illustrates the main conclusion of this passage?

A. The naming system based on clinically important features causes problems in scientific literature.
B. Changing the naming system would allow faster progress to be made in scientific medical research.
C. The naming system should be changed to reflect the features of body parts which are most important scientifically.
D. The current naming system is sufficient and should not be changed to help lazy scientists who cannot be bothered to do fact-checking.
E. It is more important to have good doctor-patient relations than good progress in scientific research.

Question 35:

Applicants for language teacher training have to specify which languages they studied as part of their degree. 180 people applied for teacher training. Of these, 128 did French as part of their degree. Half as many as did French did Spanish. Three quarters as many as did Spanish did not do either French or Spanish.

How many must have done both French and Spanish?

A. 12 B. 24 C. 36 D. 48 E. 60

Question 36:

"The government has recently had some impressive results from a campaign to reduce drug usage via education about the effects and harm that drugs can cause. However, this strategy fails to tackle one of the major causes of drug usage, namely social deprivation. Many people from deprived backgrounds take drugs due to peer pressure, as they feel that everyone around them is doing it, and that if they do not partake in the drug taking as well, they will be shunned by those around them. Here, no amount of education of the negative side effects of drugs can persuade them not to take them, because the simple fact is that in this situation there are also negative side effects to *not* take drugs, namely social exclusion. Thus the only way to persuade these people to not take drugs is to put more funding into youth centres and recreational activities, thus providing an alternative outlet to these young people, by which they can escape the social exclusion caused by not taking drugs, and engage with a different set of people, who also do not wish to take drugs. Thus, education alone cannot provide any further reductions in drug usage."

Which of the following would most weaken this argument?

A. France has put significantly more money into tackling drugs than the UK over the last few years, but also focuses on education about the negative side of drugs. France has achieved roughly the same reduction in drug usage as the UK thanks to the campaigns.
B. Germany has recently increased spending by over 50% on youth centres and sports groups in areas of high social deprivation, but has had no more success than the UK in reducing drug usage.
C. Countries which do not have significant social deprivation tend to have little issue with drug usage.
D. Ireland had a significant drug problem 15 years ago, which was vastly more challenging than the UK's current drug problem. The Irish government put significant funding into providing sports facilities in areas of high social deprivation, and achieved a remarkable reduction in drug usage.
E. 7 years ago, the Spanish government initiated a wide-reaching campaign, with significant funding to provide education about drugs in primary schools, teaching young children how drugs could damage people's lives. Since the launching of this programme, Spain has seen a reduction in drug usage which far exceeds that seen in the UK.

Question 37:

A touring musical goes to 9 different locations every two months and spends the same number of days being performed in each place. It takes 2 full days to travel between places, i.e. if the last performance in Newcastle is on Tuesday, the musical's first performance in its new location, York, can be at the earliest on Friday. The show is performed every evening, and there are additional matinee (afternoon) performances every Monday, Wednesday and Saturday. Given 1000 people attend each performance, what is the maximum number of people that can see the show during June and July combined in any year?

A. 61,000 B. 64,000 C. 70,000 D. 72,000 E. 80,000

Question 38:

A swimming tournament involves each of 72 competitors doing 3 timed heats. The 216 swims are ranked (i.e. the fastest swim is "1", the second fastest is "2"...) and the ranks of each swimmer's 3 swims are added together to give that swimmer an overall score, i.e.. if Jess swims the 10th, 20th and 50th fastest swims, her score is 80. Swimmers with total scores of 80 or less proceed to the final. If every swimmer swims exactly the same time in each of their swims, and no swimmer swims the same time as any other swimmer, how many swimmers will proceed to the final?

A. 9   B. 10   C. 11   D. 12   E. 13

Question 39:

"The damage that has been wrought on the environment by the use of fossil fuels is well established, and it is widely accepted as essential that we pursue other means of energy production. However, in spite of this, one of the major alternatives to fossil fuels, Nuclear Power, has met with fierce criticism from the public in Europe, most of whom are fearful of it. Some point to nuclear disasters such as Chernobyl and Fukushima as evidence that Nuclear power is not safe, and should not be widely implemented. However, these widely-held fears are not logical, and Nuclear power is perfectly safe. The two disasters can easily be explained: at Chernobyl a misguided procedure was carried out, during which some essential safety protocols were not followed; and the Fukushima disaster was caused by an earthquake and a tidal wave of a magnitude never seen in Europe."

Which of the following <u>cannot</u> be reliably concluded from this passage?

A.   The widespread fear of Nuclear Power in Europe is not logical.
B.   The nuclear disasters of Chernobyl and Fukushima have caused the widespread fear of Nuclear Power in Europe.
C.   The Fukushima nuclear disaster was caused by a tidal wave and an earthquake.
D.   Fossil fuels are well established to cause environmental damage.
E.   Nuclear Power is perfectly safe.

Question 40:

Bottles of squash contain 1 litre to the nearest decilitre. Megan wants to make 12 litres of a dilute squash mixture which contains exactly 3 times as much water as squash. How many bottles of squash does she need to buy to guarantee she has enough squash?

A. 2   B. 3   C. 4   D. 5   E. 6

Question 41:

Alice and Andrew both go for a run on Sunday morning. In 1 hour, Alice runs 4 times around the park and Andrew runs 3 times around the park. Amanda can only run at 60% of the speed that Alice can. Next week, Andrew and Amanda both run a 10km race. Amanda runs the race at a speed of 200 metres per minute. Assuming they both run at the same constant speed they do in training, how long in minutes and seconds does Andrew take to run the race?

A. 40:00   B. 42:30   C. 44:50   D. 46:00   E. 48:20

Question 42:

"The UK public should accept an increase in their national insurance contributions. Nobody wants to pay more taxes, but it is a fact that the population is ageing. Elderly people are more prone to health problems, meaning that the NHS requires more funds to deal with this extra workload. Given the choice, the vast majority of the people in Britain would choose having a properly funded NHS over having lower taxes."

Which of the following best illustrates the main conclusion of this passage?

A.   Nobody wants to pay more taxes.
B.   The vast majority of the public would choose having a properly funded NHS over having lower taxes.
C.   Elderly people have more health problems.
D.   The UK has an ageing population.
E.   The UK public should accept an increase in their national insurance contributions.

Question 43:

	Pool A	Pool B	Pool C	Pool D
1st	France	Argentina	England	South Africa
2nd	Holland	Mexico	Nigeria	Brazil
3rd	United States	Denmark	Germany	Japan
4th	India	Korea	Ghana	Algeria
5th	Australia	Switzerland	Portugal	Serbia
6th	Greece	New Zealand	Honduras	Uruguay
7th	Chile	Slovakia	Cameroon	Paraguay

The table above shows the final standings in the pool stages of a football competition. The top 2 teams from each pool progress into the quarter-finals. The fixtures for the quarter-finals are determined as follows:

QF1: Winners Pool A vs. Runner up Pool B   QF3: Winners Pool C vs. Runner up Pool D
QF2: Winners Pool B vs. Runner up Pool C   QF4: Winners Pool D vs. Runner up Pool A

The winners of QF1 then play the winners of QF3 in one semi-final, and the winners of QF2 and winners of QF4 play each other in the other semi-final. The winners of the semi-finals progress to the final.

Which of these teams could England play in the final?
A. Nigeria   B. France   C. Mexico   D. Denmark   E. Brazil

Question 44:

Kelly is working on a school project. She begins by making a large card surface on which to display her project by attaching A4 (300mm x 210mm) sheets of card together with tape. She applies tape to the entire back and front of every join she makes. She wants her large card to be at least 1 metre by 1 metre and doesn't cut or overlap any card.

How much tape does she need in total to make her large card surface?
A. 3.51m   B. 5.85m   C. 7.02m   D. 10.7m   E. 107m

Question 45:

Katie's netball team have played 24 matches this season. They play each team once at home and once away. In total they have won 18 matches. They have won twice as many matches at home as away. They have not drawn any matches. How many more matches have they lost away than lost at home?
A. 2   B. 3   C. 4   D. 5   E. 6

Question 46:

"Science fiction movies have a lot to answer for. In the UK, many Sci-fi movies have contributed to a remarkably poor understanding of science, and in some cases, a misguided and illogical fear of scientific progress. One prime example is the genre of zombie horror, which frequently features a virus taking over people, modifying their behaviour in a matter of seconds, and turning them into deranged, mindless zombies. Such a phenomenon is impossible; the closest any virus comes is the rabies virus, which takes several days at the minimum to stimulate aggressive, raging behaviour. Such illogical fear often leads to less public support for science funding, which severely hinders research into cures for many devastating diseases. Thus, it is clear that without science fiction movies, many deaths could be prevented."

Which of the following best illustrates a flaw in this passage?
A. The example of zombie horror movies may not be applicable to other sci-fi movies.
B. There is no evidence that any viruses have the ability to infect people and modify their behaviour within a few seconds.
C. It has assumed that if research on diseases were not hindered, many deaths could be prevented.
D. It has assumed that there are no other contributing factors to public fear of scientific progress.
E. It has assumed that the public fear of science is illogical.

Question 47:

Ashley, Ben, Callum, Dave and Ed agree to meet in the city centre at 1pm. Ashley walks from his house 8km away at a speed of 8km per hour. Ben gets the bus, which takes 40 minutes and departs from his house at 25 and 55 minutes past the hour, every hour. Callum cycles 12.5km at a speed of 12km per hour. Dave gets the train, which goes every 10 minutes and takes 20 minutes, but he has to walk to the station which takes 25 minutes. Ed drives to the park and ride which takes 10 minutes, then gets the park and ride bus which comes every 10 minutes and takes 15 minutes.

Who has to leave their house the earliest to get to the city centre on time?

A. Ashley   B. Ben   C. Callum   D. Dave   E. Ed

Question 48:

6 friends from university all send each other Christmas cards. Posting each card costs £0.50, apart from cards to and from Sophie, who lives abroad. Posting cards abroad costs £1.50, and sending cards from abroad costs £1.20.

How much in total is spent by the 6 friends on sending cards?

A. Between £20 and £21.99   D. Between £26 and £27.99
B. Between £22 and £23.99   E. Between £28 and £29.99
C. Between £24 and £25.99

Question 49:

	Goals scored For	Goals scored Against
City	10	4
United	8	5
Rovers	1	10

The table above shows the goal scoring record of teams in a football tournament. Each team plays the other teams twice, once at home and once away. Here are the results of the first 4 matches:

United 2 – 2 City   City 2 – 1 Rovers
Rovers 0 – 3 City   Rovers 0 – 3 United

What were the results of the final two fixtures?

A. United 2 – 0 Rovers, City 0 – 0 United   D. United 1 – 0 Rovers, City 2 – 2 United
B. United 1 – 0 Rovers, City 1 – 1 United   E. United 2 – 0 Rovers, City 3 – 1 United
C. United 0 – 0 Rovers, City 2 – 1 United

Question 50:

"The UK has recently had an election, and the opposition won, meaning a new cabinet has been formed. The new education minister has argued that the UK should introduce more coursework into GCSE and A Level courses, because it is a much fairer way to assess students than via examination. In fact, coursework does not measure a student's ability in a subject, but how much help they receive outside their lessons. Students with help from relatives or friends who are knowledgeable in certain subjects are shown to perform much more strongly than others with a similar ability but less help. Thus, coursework is clearly not a fairer method of assessment than examinations, so the new education minister is clearly incorrect."

Which of the following best illustrates the main conclusion of this passage?

A. The new education minister is not a logical person.
B. Coursework is not the fairest method of assessment for GCSE and A Level courses.
C. Coursework measures students on the level of outside help they receive, not their ability in a subject.
D. The education minister's argument is incorrect.
E. There are no methods of assessment fairer than examination.

END OF SECTION

YOU MUST ANSWER <u>ONLY</u> <u>ONE</u> OF THE FOLLOWING QUESTIONS

Question 1:
To what extent are 'logical' and 'rational' synonymous?

Question 2:
In what instances is aggression justified?

Question 3:
What are the limits of scientific theories of human behaviour?

Question 4:
Assuming time travel was possible, could we learn more from the past or the future?

END OF TEST

Mock Paper D

Question 1:

"Irish Folk Band, the Willow, have recently signed a contract with a new manager, and are organising a new musical tour. They and their manager are discussing which country would be best to organise their tour in. The lead singer of the Willow would like to organise a tour in Germany, which has a rich history of folk music. However, the manager finds that ticket sales for folk music concerts in Germany have been steadily declining for several years, whilst France has recently seen a significant increase in ticket sales for folk music concerts. The manager says that this means the group's ticket sales would be higher if they organise a tour in France, than if they organise one in Germany."

Which of the following is an assumption that the manager has made?

A. The band should prioritise profits and organise a tour in the most profitable country possible.

B. The band should not embark upon a new tour and should instead focus on record sales.

C. The decrease of ticket sales in Germany and the increase in France means that the band will sell fewer tickets in Germany than in France.

D. There will not be other countries which are even more profitable than France to organise the tour in.

E. Folk music is popular in France.

Question 2:

"Tom and Rob are going on a trekking holiday in Africa, organised by Wild Africa Inc. The holiday will include hiking, a safari day, kayaking and mountain climbing. Tom suffers from Asthma, and the pair does some research to find out if this will affect any part of their holiday. Rob is disappointed to find the following clause in the terms and conditions of Wild Africa Inc:

5.2 Customers suffering from severe Asthma, Breathing difficulties, Anaphylaxis or heart conditions are not allowed to take part in mountain-related activities on holidays organised by Wild Africa Inc.

Rob concludes that this means that Tom will not be allowed to take part in the mountain climbing section of their holiday."

Which of the following is an assumption that Rob has made?

A. The clause rob has found applies to their holiday.

B. Tom was intending to take part in the mountain climbing section of the holiday.

C. If Tom is not able to take part in the mountain climbing, Rob will not be able to either.

D. Tom suffers from severe asthma.

E. Wild Africa Inc. is following the correct health and safety regulations.

Question 3:

"Grace and Rose have both been attending an afterschool gymnastics class, which finishes at 5pm. After the class has finished, Grace and Rose cool down and change out of their gym clothes before heading home. Both girls depart at 5:15pm. Grace and Rose both live a 1.5 mile walk away from the local gymnasium. Therefore, they will definitely arrive home at the same time."

Which of the following is NOT an assumption made in this argument?

A. Both girls will walk at the same speed.

B. Both girls departed at the same time.

C. The gymnastics class is being held at the local gymnasium.

D. Grace will not get lost on the way home.

E. Both girls are walking home.

Question 4:

"John is a train enthusiast, who has been studying the directions in which trains travel after departing from various London Stations. He finds that trains departing from King's cross station in London head North on the East Coast Mainline, and travel to Edinburgh. Trains departing from Waterloo Station head West on the Southwest Mainline and travel to Plymouth. Trains departing from Victoria Station head South and travel to Kent.

John surmises that presently, in order to travel on a train from London to Edinburgh, he must get on at King's Cross Station."

Which of the following is an assumption that John has made?

A. The East Coast mainline has the fastest trains.
B. It would not be quicker to take a train from Waterloo to Southampton Airport, then travel to Edinburgh on an aeroplane.
C. Rail lines will not be built that will allow trains to travel from Waterloo Station or Victoria Station to Edinburgh.
D. Trains departing from King's Cross do not have any other destinations other than Edinburgh.
E. There are no other train stations in London from which trains may travel to Edinburgh.

Question 5:

"Modern cars have a many safety features such as seatbelts, crumple zones and enhanced braking systems. Though done in the name of safety, this gives many drivers a false sense of security, and causes them to drive much faster. This increase in driving speeds results in a great number of crashes at high speed, causing thousands of deaths each year as no amount of safety features is able to prevent deaths in the face of the immense forces involved in a crash at high speed. Therefore, we should install a new feature in cars, a spike on the steering wheel, pointing directly at the driver's chest. This would save thousands of lives each year, as when driving at lower speeds, drivers have much more control, and crashes are much more readily avoided."

Which of the following is an assumption made in this argument?

A. A spike in the steering wheel pointing at the driver's chest would cause people to drive more slowly.
B. Improving safety features in cars would not be sufficient to reduce deaths from high-speed crashes.
C. Crashes are much more readily avoided when driving at lower speeds.
D. Safety features have caused a lot of high-speed crashes.
E. High-speed crashes are causing millions of deaths each year.

Question 6:

"Tanks and armoured vehicles were a hugely influential factor in all battles in World War 2. German tanks were highly superior to the tanks used by France, and this was an essential reason why Germany was able to defeat France in 1940. However, Germany was later defeated in World War 2 by the Soviet Union. Germany lost a number of key battles such as the Battle of Stalingrad and the Battle of Kursk. These victories were essential for the eventual victory of the Soviet Union over Germany. Therefore, the Soviet Union's tanks in the battles of Stalingrad and Kursk must have been superior to those of Germany."

Which of the following is an assumption made in this argument?

A. Tanks were hugely influential in the Battle of Stalingrad.
B. The Battles of Stalingrad and Kursk were essential for the Soviet Union's victory over Germany.
C. The reasons why the Soviet Union defeated Germany in battle were the same as the reasons why Germany defeated France in battle.
D. German tanks being superior to those used by France was an essential reason why Germany was able to defeat France.
E. If the Soviet Union's tanks were superior to Germany's tanks, the Soviet Union's armoured vehicles must also have been superior to Germany's armoured vehicles.

Question 7:

"In the Battle of Waterloo, in 1815, French Emperor Napoleon Bonaparte's army was defeated by a British army commanded by British General Arthur Wellesley, Duke of Wellington. Essential to The British army's victory was the arrival of a group of Prussian reinforcements led by Field Marshal Von Blucher, which joined up with The British army and allowed them to overwhelm Bonaparte's left flank. Bonaparte had been aware of the threat posed by Von Blucher's Prussians, and had detached a force of French soldiers several days earlier under the command of Field Marshal Grouchy, with orders to engage the Prussians led by Von Blucher, and prevent them joining up with The British Army. However, whilst dining at a local inn, Grouchy mistook the sounds of gunfire for thunder, and believed that the battle had been cancelled. He therefore disobeyed his orders and did not engage the Prussians commanded by Von Blucher. Therefore, if Field Marshal Grouchy had not made this mistake and had engaged the Prussian force as commanded, the British would not have won the Battle of waterloo."

Which is the best statement of a flaw in this argument?

A. It implies that Field Marshal Grouchy was an incompetent commander, when in fact he was a highly respected general of the day.
B. It assumes that had Grouchy engaged the Prussian force, he would have been able to successfully prevent them joining up with the British army.
C. It assumes that the British army would not have been victorious without the arrival of the Prussian reinforcements.
D. It ignores the other mistakes made by Napoleon which contributed to the British army being victorious in the Battle of Waterloo.
E. It implies that thunder and gunshot sounds are frequently mistaken by generals.

Question 8:

"A cruise ship is sailing from Southampton to Barcelona, making several stops along the way at Calais and Bordeux, in France, Bilbao in Spain, and Porto in Portugal. At each stop, the ship must wait in a queue to be assigned a dock at which it can pull in, refuel and resupply. The busier the port, the longer the ship will have to queue to be assigned a dock. The Captain of the ship is planning the journey, and knows he must work out which ports will have the longest queues.

The Captain made the same journey last year, and found out that Bilbao was the busiest port in Europe during the course of the journey. He also knows that Bordeux is the busiest port in France, and that Porto is the busiest port in Portugal. Whilst he is planning the journey, he discovers that Calais is busier than Porto. The Captain concludes that he must plan for Bilbao to have the longest queue in the journey, Bordeux to have the second longest queue, Calais to have the third longest queue, and Porto to have the fourth longest queue."

Which of the following best illustrates a flaw in the Captain's reasoning?

A. Porto is less busy than Calais, but may be busier than Bordeux.
B. The rankings may have changed and Bilbao may no longer be the busiest port in Europe.
C. Just because a port is busier does not necessarily mean it will have the longest queues.
D. The ship may not have time to make all the stops.
E. The captain has forgotten to consider how many passengers will embark and disembark at each stop.

Question 9:

"Every year in Britain, thousands of children need to be taken to hospital due to unexpected Asthma attacks. Many of these attacks could be treated by inhalers, and not require hospitalisation, if inhalers were quickly available. However, in spite of this problem, very few schools stock inhalers. If the government enacted legislation requiring schools to keep a stock of inhalers, then many of these hospitalisations could be avoided."

Which of the following best illustrates a flaw in the reasoning of this argument?

A. It ignores the logistical problems associated with stocking inhalers in schools.

B. It assumes that inhalers could prevent asthma attacks as well as treating them.

C. It ignores the fact that inhalers can sometimes cause other medical problems due to allergic reactions.

D. It implies that if inhalers are stocked in schools they will be quickly available to children, in time to treat unexpected asthma attacks.

E. It neglects to consider asthma attacks which cannot be treated by an inhaler.

Question 10:

"The aeroplane was a marvel of modern engineering when it was first developed in the early 20th Century, and was testament to human ingenuity. Throughout the 20th Century, the aeroplane allowed humans to travel more freely and widely than ever before, and allowed people to see and appreciate the stunning natural beauty that the world has to offer. However, aeroplanes also produce lots of pollution, such as Carbon Dioxide and Sulphur Oxide. High levels of Carbon Dioxide in the atmosphere are currently causing global warming, which is destroying or damaging many natural environments throughout the world.

Therefore it is clear that the aeroplane, which once offered such opportunity to appreciate the world's natural beauty, has been largely responsible for damage to various natural environments throughout the world. We must now seek to curb air traffic in order to save the world's remaining natural environments."

Which of the following is the best statement of a flaw in this argument?

A. It assumes that aeroplanes are a major reason for the high levels of Carbon Dioxide in the atmosphere which are currently causing global warming.

B. It assumes that aeroplanes offer greater opportunity to appreciate the world's natural environments.

C. It assumes that high levels of Carbon Dioxide are responsible for global warming.

D. It does not consider the effects of Sulphur Dioxide pollution released by aeroplanes.

E. It implies that we should take action to prevent damage to the world's natural environments.

Question 11:

"Professors from the Department of Pathology at Amazonia University are conducting research into possible new treatments for malaria, which is caused by a microbe known as Plasmodium. Research from Sierra Leone, a third world country with a high rate of malaria, has found that liver cells in malaria patients are reactive to the antibody Tarpulin. Plasmodium is known to infect liver cells, and thus liver cells would react to Tarpulin if Plasmodium itself was reactive to Tarpulin. Thus the professors at Amazonia University begin to research how Tarpulin can be used to target Plasmodium and treat malaria. However, this research will not be successful, because liver cells would also react to Tarpulin if the wrong solution is used whilst conducting the experiments. Since malaria is not prevalent in Amazonia, the professors must rely on the data from Sierra Leone. If the experiments in Sierra Leone used the wrong solutions, then the liver cells would react to Tarpulin even if Plasmodium does not react to Tarpulin."

Which of the following best illustrates a flaw in this argument?

A. From the fact that Plasmodium infects liver cells, it cannot be inferred that infected liver cells would react to Tarpulin if Plasmodium does.

B. From the fact that the research was carried out in Sierra Leone, it cannot be inferred that the wrong solutions were used.

C. From the fact that the wrong solutions are used, it cannot be inferred that the liver cells would react to Tarpulin.

D. From the fact that Plasmodium is reactive to Tarpulin it cannot be assumed that Tarpulin can be used to combat Plasmodium.

E. From the fact that Liver cells react to Tarpulin, it cannot be inferred that Plasmodium is reactive to Tarpulin.

Question 12:

"Ancient Egypt was one of the world's most powerful nations for several thousand years, and built wondrous structures including the Sphinxes and the Great Pyramids to serve as a permanent reminder of its stature. Many other powerful nations throughout the ages have also built magnificent structures, such as the Colosseum built by the Romans, the Hanging Gardens of Babylon built by the Persians and the Great Wall of China built by the Chinese. As well as building magnificent structures, Rome, Persia and China had one other thing in common, namely a very strong military. Thus, history clearly shows us that in ancient times, for a nation to be a powerful nation, it must have had a very strong military. In addition to building great structures such as the pyramids, Ancient Egypt must have also possessed a very strong military."

Which of the following best illustrates the main conclusion of this argument?
A. In order to be a powerful nation, a nation must build magnificent structures.
B. In ancient times a very strong military was required to be a powerful nation.
C. Ancient Egypt built magnificent structures; therefore it must have been a powerful nation.
D. Rome, Persia and China were all powerful nations.
E. Ancient Egypt was a powerful nation; therefore, it must have had a very strong military.

Question 13:

Global warming is widely presented in modern society as a cause for significant concern. One particular area often thought to be at risk is the ice caps of the North and South Poles, which are often presented to be at risk of melting due to increased temperature. Environmentalist groups often campaign for energy consumption to be reduced, thus reducing CO_2 emissions, the leading cause of global warming. However, recent research shows that the North and South Poles are actually becoming cooler, not warmer, thanks to mysterious and unexplained weather patterns. Clearly, high energy consumption is not contributing to damage to the polar ice caps.

Which of the following best states a conclusion that can be drawn from this argument of this argument?
A. There is no point in reducing energy consumption for environmental reasons.
B. Reducing energy consumption will not reduce CO_2 emissions.
C. We should trust the recent research stating that the North and South poles are becoming cooler.
D. Reducing energy consumption will not contribute to saving the polar ice caps.
E. We should not be concerned about damage to the Polar Ice caps.

Question 14:

"In 1957 the drug Thalidomide was released, and used to relieve nausea and morning sickness during pregnancy. The pharmaceutical company which released Thalidomide had carried out extensive testing of the drug, and had carried out more tests than was required for new drugs in the 1950s. No adverse affects were reported, and the drug was thought to be safe and effective. However, after it was released, Thalidomide was found to be responsible for severe deformities in thousands of babies whose mothers had taken the drug whilst pregnant with them. When further research was carried out, it was found that the molecules in Thalidomide could adopt 2 molecular structures, known as Isotopes. One of these Isotopes was perfectly safe, but the other caused significant biological problems in pregnant women and had been responsible for the deformities in the babies. The company producing Thalidomide had not been aware of this 2nd Isotope when developing the drug."

Which of the following is a conclusion that can be drawn from this passage?
A. The company producing Thalidomide had acted irresponsibly by not carrying out the required level of testing for the drug.
B. No Isotopes of Thalidomide are safe.
C. The drug testing requirements in 1950s were not sufficient to identify all possible Isotopes of a given drug.
D. Thalidomide was not effective at relieving nausea and morning sickness.
E. The dangerous Isotope of Thalidomide was not effective at relieving nausea and morning sickness.

Question 15:

"As winter draws in and the weather turns colder, many bird species struggle to find enough places to live in modern cities, and populations are at risk of declining. The RSPB is working to tackle the problem, and has pledged £500,000 towards building purpose-built homes for birds throughout many of the UK's major cities, with the aim of keeping bird populations as high as possible. However, the RSPB is taking a very narrow view of conservationism. The amount of money they have pledged should instead be used to purchase land and build mini nature reserves within major cities. Estimates suggest this would help a similar number of birds as the planned purpose-built homes. It would also offer the added benefit of helping many other types of wildlife, not just birds."

Which of the following best illustrates the main conclusion of the above passage?
A. The RSPB should seek to help as much wildlife as possible, not just birds.
B. The purpose-built homes for birds in major cities are a waste of money.
C. The estimates that building nature reserves would help as many birds as the planned purpose-built homes are correct.
D. The RSPB should build nature reserves instead of purpose-built homes in order to help the maximum amount of wildlife possible.
E. Nature reserved would help more wildlife than purpose-built homes for birds.

Question 16:

"The release of CO_2 from consumption of fossil fuels is the main reason behind global warming, which is causing significant damage to many natural environments throughout the world. One significant source of CO_2 emissions is cars, which release CO_2 as they use up petrol. In order to tackle this problem, many car companies have begun to design cars with engines that do not use as much petrol. However, engines which use less petrol are not as powerful, and less powerful cars are not attractive to the public. If a car company produces cars which are not attractive to the public, they will not be profitable."

Which of the following best illustrates the main conclusion of this argument?
A. Car companies which produce cars that use less petrol will not be profitable.
B. The public prefer more powerful cars.
C. Car companies should prioritise profits over helping the environment.
D. Car companies should seek to produce engines that use less petrol but are still just as powerful.
E. The public are not interested in helping the environment.

Question 17:

"Penicillin is one of the major success stories of modern medicine. Since its discovery in 1928, it has grown to become a crucial foundation of medicine, saving countless lives and introducing the age of antibiotics. Alexander Fleming is today given most of the credit for introducing and developing antibiotics, but in fact Fleming played a relatively minor role. Fleming initially discovered Penicillin, but was unable to demonstrate its clinical effectiveness, or discern ways of reliably and consistently producing it. 2 other scientists called Howard Florey and Ernst Chain were actually responsible for developing Penicillin to the point where it could be reliably produced and used in medicine, to treat infections in patients. Clearly, the credit for the wonders worked by Penicillin should not go to Fleming, but to Florey and Chain."

Which of the following best illustrates the main conclusion of this argument?
A. Fleming was unable to develop penicillin to the point of being a viable medical treatment.
B. The credit for Penicillin's effects on medicine should go to Ernst Chain and Howard Florey, not to Alexander Fleming.
C. Without Chain and Florey, Penicillin would not have been developed into a viable treatment.
D. Alexander Fleming only played a small role in the process of Penicillin becoming a feature of modern medicine.
E. Alexander Fleming is not given enough credit for his role in the development of penicillin.

Question 18:

"Barns and Co. wholesalers are seeking ways to increase staff productivity, and have commissioned research into methods that they can use to do this. One of their researchers reports that greater teamwork and co-operation between staff is well proven to increase productivity in businesses. She argues that group days out for staff, doing activities such as paintballing or boating, are proven to instil greater teamwork in staff.

One of the board members raises an objection, claiming that the expense and lost working days stemming from organising a staff day out will cause an immediate drop in productivity. The researcher states that this is indeed a downside to organising a staff day out, but argues that in the long term the increase in productivity from organising such an event will more than account for this short term drop."

What of the following is the best illustration of the main conclusion of the researcher's argument?

A. Barns and Co. should not organise a staff day out due to the immediate negative effect on productivity this will have.

B. The board should ignore the short term decrease in productivity that will result from organising a staff day out.

C. Barns and Co. must organise a staff day out or they risk a decrease in productivity.

D. The board member who questioned her research should be sacked.

E. The board of Barns and Co. should accept a short term decrease in productivity from organising a staff day out, in order to achieve the long-term increase in productivity that will result.

Question 19:

"Worcestershire Aquatic Centre is seeking to recruit a new dolphin trainer. They interview several candidates, and find that there are 2 candidates which are clearly more suitable than the others. They give both of these candidates a 2nd interview, with further questions about their experience and qualifications.

They discern that Candidate 1 has a proven capability to perform well to crowds, which is likely to bring in more profit to the Aquatic Centre as more people will come and watch the dolphin shows. However, Candidate 2 has more experience at handling dolphins, and a proven ability to do this successfully. The manager of the aquatic centre tells the recruiting officer that it is more important that the new dolphin trainer is able to successfully handle the dolphins, than that they are able to increase profit."

Which of the following best illustrates a conclusion that can be drawn from the above passage?

A. The recruiting officer will should hire Candidate 1 in order to increase profits at the Worcestershire Aquatic Centre.

B. The recruiting officer should ignore the manager, and hire whichever candidate he feels is the best.

C. The recruiting officer should hire Candidate 2, to ensure that the new trainer will be able to successfully handle the dolphins.

D. The recruitment officer should hire both candidates.

E. The 2 candidates are no more suitable than the others which were interviewed, and neither should be hired.

Question 20:

"Thrills and Spills Inc. is a theme park and water park in England. The park has Rollercoasters, Water Slides, Fun Rides, and Amusements. Currently, the park charges a fixed entry fee, and after customers have paid this they are able to access all rides free of charge. Stephen thinks this is unfair, and argues that it should be replaced by a system in which customers only pay for the rides you go on. Thus, only users who use Rollercoasters should pay for access to Rollercoasters, and likewise for the other types of ride."

Which of the following proposals does NOT follow the same reasoning that Stephen has used?

A. Every motorist is charged a flat rate of road tax, and the money raised is used towards the upkeep of all roads. This should be abolished, and replaced by a new system of toll roads. Then, only motorists who use motorways would have to pay towards their upkeep, and only motorists who use country roads would have to pay towards the upkeep of these roads.

B. Under the current system, everybody in the UK pays a flat rate of National Insurance tax, which is used to fund the NHS. This is unfair, and should be replaced by a system in which everybody pays for only the services that they use. For example, only those who require X-Rays should have to pay for the upkeep of the X-Ray departments in the NHS.

C. Communism is a terrible system of government, in that everybody has to pay for all public services regardless of which ones they use. A much better system is to privatise all public services, and make users pay, so that only people who use each service should pay for it.

D. Currently, everybody who watches TV pays a Television license, regardless of which channels they watch. This is unfair, and should be replaced by making all channels pay-to-view, so customers only have to pay for the channels they watch.

E. Currently everybody who uses the London Underground pays a flat rate for a day's access to the tube, regardless of how many journeys they make. This should be replaced by a system in which customers pay for each journey, and thus pay an amount dependent on how much they use the Tube.

Question 21:

"There has recently been a new election in the UK, and the new government is pondering what policy to adopt on the railway system in the UK. The Chancellor argues that the best policy is to have an entirely privatised railway system, which will encourage different train companies to be competitive, and try and attract customers by providing the best service at the lowest price, thus driving down costs and increasing quality for customers. However, the Transport Minister argues that this is a short-sighted policy. She argues that privatised companies will only run services on the most profitable lines, where there are lots of passengers. Under this system, train companies may choose not to run many services to rural areas. This will lead to rural communities being cut off and will cause issues of segregation and a lack of opportunities for people in these communities. She argues that public funding should be put towards rail services in order to ensure that people in rural communities are adequately served by rail services."

Which of the following, if true, would most strengthen the Transport Minister's argument?

A. The Transport Minister has ultimate power over railway policy, and she can overrule the Chancellor if she sees fit.

B. Many train services to rural communities currently have low passenger numbers, and are unlikely to be profitable.

C. French rail services receive high level of public funding, and users of these services enjoy good quality and low prices.

D. American railway services are privatised with no public funding, and yet rural communities in America are well served by railway services.

E. The Prime Minister agrees with the Transport Minister's line of argument. He sympathises with rural communities and does not believe in a privatised rail system.

Question 22:

"AIDEL supermarket is seeking to attract more customers, and gain a greater share of the local market. The board know that if they lower their prices, they will attract more customers. The board decides to lower prices across the store. Therefore they can expect their customer numbers to increase."

Which of the following most closely follows the reasoning in this paragraph?

A. Nisha is flying to America from London. If she flies from Gatwick airport, the price of the flight will be less than if she flies from Heathrow airport. Therefore, if she wishes to fly at a cheaper price than it will cost to fly from Heathrow, she must fly from Gatwick.

B. If a train is to arrive into Middlesbrough Station, it must first go through Thornaby Rail Station. At 18:53, a train pulls into Middlesbrough Station. Therefore, we know it must have passed through Thornaby Rail Station before 18:53.

C. Richard decided a long time ago that if he ever travels to London, he will ensure to go and watch Phantom of the Opera whilst he is there. For his 20th Birthday, Richard's parents pay for him to travel to London and spend several days there. Therefore, Richard will go and see Phantom of the Opera.

D. Red Squirrels are currently on the decline in the UK, mostly due to competition from Grey Squirrels. If the population of Red Squirrels is to increase, the number of Grey Squirrels must first decrease. The population of Grey Squirrels has recently decreased, and therefore we can expect to see an increase in the population of Red Squirrels.

E. If Norwich Rugby team are to win the East-of-England League, they must defeat Cambridge. Norwich are crowned champions of the East-of-England League. Therefore Norwich must have defeated Cambridge.

Question 23:

"The United States of America has a healthcare system which works on a pay-per-use basis, where users pay a certain amount dependent on how much they use the health service. This is ethically outrageous, as many citizens are too poor to pay for healthcare, and cannot therefore receive the treatment they need. It is much better to have a system where everybody pays an amount depending on how wealthy they are, such that the rich would pay more than the poor. This should then be used to fund the healthcare service, and treatment then provided to users based on need, and not on whether they can afford to pay for it."

Which of the following best illustrates the principle used in this argument?

A. There is shocking differentiation in the wealth of different people living in the world. Some live lives of vast extravagance, whilst others do not have the means to buy essentials. Therefore, we should tax the wealthy, and use the money to fund support for the poor.

B. The Police should not charge people who require help and assistance. Instead, everybody should pay a fixed amount, and the money used to fund the Police. The Police should then provide help and assistance to people based on need, not wealth.

C. Healthcare should be a universal right, so people should not have to pay for healthcare which they use. Instead, the health service should be paid for by taxing those responsible for major health conditions. For example, fast food chains should be taxed, as their products promote obesity, a major health concern. The Health Service should then provide care based on need, not wealth.

D. The Police should not be a publically funded body. Instead, those who require help from the police should pay for it, depending on how much help they require. This would encourage people to be less dependent on the police

E. The Fire Service should not charge people who require help with house fires. Instead, everybody should pay an amount depending on their income/personal wealth, and the money used to fund the Fire Service. The Fire Service should then provide help to people based on need, not wealth.

Question 24:

"The North York Moors is one of several National Parks in England. The Management team has been awarded a grant from the National Lottery and is looking for a way to attract more visitors to the Moors. Sam suggests that they invest in enhancing the natural landscapes present in the Moors, thus creating more beauty, and making people more inspired to visit. However, Lucy disagrees, and feels that they should invest in more visitor centres and information points. Lucy's argument is that whilst this will be more costly in terms of staffing these centres, the increase in visitor numbers will bring in more income for the Moors, and will counteract this extra cost."

Which of the following, if true, would most weaken Lucy's argument?

A. Visitor centres in other national parks do not generally generate as much revenue as they cost to staff.
B. National Lottery grants have a history of being badly spent by National Parks such as the North York Moors.
C. There are large numbers of people who are interested in volunteering to help the North York Moors and would be happy to staff visitor centres.
D. Another National Park, the Yorkshire Dales, has recently opened up 5 new visitor centres and seen their profits increase significantly.
E. The North York Moors is currently struggling to attract visitors.

Question 25:

"Recent statistics have found that in many important areas of current debate, the public are consistently incorrect in what they believe is happening. For example, the public believe that 24% of Benefit claims are fraudulent, when in fact 0.7% of benefit claims are fraudulent. The public also believe that Violent Crime has increased by 10% in the last 5 years, when it has actually fallen by 25%. Finally, the public believe that 5 times more money is spent on Jobseeker's allowance than on Pensions, when in fact 15 times more money is spent on Pensions than Jobseeker's allowance. Therefore, efforts to engage with the public on these issues of debate should be abandoned as it is pointless to try and argue with people who are so clearly wrong about the facts."

Which of the following best illustrates the principle used in this argument?

A. We should make efforts to increase public education, in order to ensure that the public are well informed about important matters.
B. It is important to ensure that the values of democracy are upheld. Therefore, when engaging on matters of medical policy, we should talk to all parties, even those which are often found to be incorrect about medical matters.
C. Kerry is looking to open up a new restaurant and is looking for advice. She should only talk to professional chefs, as these have the greatest expertise in this area.
D. There are many people who continue to campaign against vaccines, despite the fact that vaccines are one of the most beneficial discoveries of modern medicine. Research has shown time and time again that anti-vaccine campaigners are consistently wrong about how vaccines actually work. Therefore, there is no point in debating with anti-vaccine campaigners about vaccines.
E. Nobody knows everything, and everyone is incorrect on some matters. Therefore, we should seek to engage with as many people as possible on all matters.

Question 26:

When I board my train at York at 15:30, the announcer tells me it is 120 minutes to London Kings Cross. Assuming the announcement is accurate to the nearest 10 minutes and that the train is on time, what is the earliest time I might arrive at my destination which is a 10 minute walk from Kings Cross?

A. 17:20   B. 17:25   C. 17:30   D. 17:35   E. 17:40

Question 27:

My sister does 2 loads of washing per week plus an extra one for everyone who is living in the house that week. When her son is away at university, she buys a new carton of washing powder every 6 weeks, but when her son is home she has to buy a new one every 5 weeks. How many people are living in the house when her son is home?

A. 2   B. 3   C. 4   D. 5   E. 6

Question 28:

A train shuttle service runs between the city centre and the airport between 5:30am and 11:30pm on weekdays and 6:00am and midnight on weekends. There are two trains used to operate the service and each journey from the airport to the city centre or vice versa takes 24 minutes. It takes 4 minutes for the train to unload and 4 minutes to reload at the airport; 2 minutes for the train to unload and 2 minutes to reload at the city centre. What is the maximum number of journeys that can be made by the shuttle service in one day?

A. 36 B. 45 C. 60 D. 72 E. 90

Question 29:

Northern Line trains arrive into Waterloo Station every 6 minutes, Jubilee Line trains every 2.5 minutes and Bakerloo Line trains every 4 minutes. If trains from all 3 lines arrived into the station 4 minutes ago, how long will it be before they do so again?

A. 20 minutes D. 54 minutes
B. 26 minutes E. 56 minutes
C. 30 minutes

Question 30:

Sam is deciding whether to make her wedding invitations herself or get them professionally made at a cost of £1 each. She decides to work out how much it will cost to make them herself. Each invitation uses 1 sheet of cream card, 4 sheets of red paper and 1 metre of gold ribbon. She will also use a gold sticker on each invitation and stamp them with a stamper, which she will buy. The stamper needs a pad of ink which will last for 70 invitations. The table of stationery costs is shown below:

Product	Price
Red paper (pack of 100)	£2
15m roll of gold ribbon	£3
Pack of 30 gold stickers	£1
Stamper	£8
Ink pad	£4
Cream card (pack of 20)	£2

She wants to send 90 invitations and wants to have enough supplies for 4 spares only. How much will she save by making the invitations herself?

A. £15 B. £19 C. £23 D. £27 E. £31

Question 31:

Half of the boys in Mrs Nelson's class have brown eyes and two thirds of the class have brown hair. If we know that at least as many boys in the class as girls have brown hair, that there are at least as many boys than girls in the class and that there are 36 children in the class altogether, at least how many boys have both brown hair and brown eyes?

A. 2 B. 3 C. 4 D. 5 E. 6

Question 32:

Mandy is making orange squash for her daughter's birthday party. She wants to have a 300ml glass of squash for each of the 8 children attending and a 400ml glass of squash each for her and for 2 parents who are helping out. She has 600ml of the concentrated squash. What ratio of water:concentrated squash should she use in the dilution to ensure she has the right amount to go around?

A. 7:1 B. 6:1 C. 5:1 D. 4:1 E. 3:1

Question 33:

A group of students is picking A Level options from the table below. In each "block", they choose 1 option. They cannot choose the same option in more than one block, and they have to take a subject in every block. How many possible combinations of options are there?

Block A	Block B	Block C	Block D
English Literature	History	Biology	Chemistry
German	Physics	English Language	Economics
Mathematics	Psychology	Geography	Philosophy
Psychology	Spanish	Mathematics	Sociology

A. 16 B. 48 C. 64 D. 224 E. 256

Question 34:

Hannah opens a savings account for her 14 year old son who will go to university in 4 years time. She works out that he will need £19000 to live on at current prices so deposits this amount in the savings account when she opens it. The cost of living rises with inflation by 3% a year, and Hannah's money earns 4% a year on balances under £20000 or 5% a year on balances over £20000. The table below shows how much an increase of various percentages equates to for various amounts. The interest is calculated and paid into the account at the end of each year.

How much more money will Hannah's son have than he needs to live on to the nearest £100 (if he assumes the cost of living does not rise further and no further interest is earned after the start of his course in 4 years time)?

	3%	4%	5%
£19,000.00	£19,570.00	£19,760.00	£19,950.00
£19,570.00	£20,157.10	£20,352.80	£20,548.50
£19,760.00	£20,352.80	£20,550.40	£20,748.00
£19,950.00	£20,548.50	£20,748.00	£20,947.50
£20,157.10	£20,761.81	£20,963.38	£21,164.96
£20,352.80	£20,963.38	£21,166.91	£21,370.44
£20,548.50	£21,164.96	£21,370.44	£21,575.93
£20,550.40	£21,166.91	£21,372.42	£21,577.92
£20,748.00	£21,370.44	£21,577.92	£21,785.40
£20,761.81	£21,384.67	£21,592.29	£21,799.90
£21,372.42	£22,013.59	£22,227.31	£22,441.04
£21,577.92	£22,225.26	£22,441.04	£22,656.82

A. 800 B. 1000 C. 1300 D. 1400 E. 1500

Question 35:

Amaia needs to arrive at her university interview by 11am She has to travel from her house, a 10 minute walk from Southtown station, to Northtown University. Using the timetables below, what is the latest time she should she leave her house to arrive on time?

Train Timetable

Southtown	0913	0943	1013
Westtown	0924	0954	1024
Northtown West	0950	1020	1050
Northtown Central	0958	1028	1058
Easttown	1009	1039	1109

Bus Timetable

Northtown West station	959	1009	1019	1029
Northtown Shopping Centre	1009	1019	1029	1039
Northtown Central station	1020	1030	1040	1050
Northtown Football Club	1023	1033	1043	1053
Northtown University	1033	1043	1053	1103

A. 0903   B. 0913   C. 0923   D. 0933   E. 0943

Question 36:

Summer and Shaniqua are playing a game of "noughts and crosses". Each player is assigned either "noughts" (O) or "crosses" (X) and they take it in turns to choose an empty box of the 3x3 grid to put their symbol in. The winner is the first person to get a line of 3 of their symbol in any direction in the grid (vertically, horizontally or diagonally). Summer starts the game. The current position is shown below.

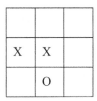

Assuming Shaniqua now plays her symbol in the square which will stop Summer being able to win the game straight away, Summer should play in either of which 2 boxes to ensure she is able to win the game on the next turn no matter what Shaniqua does?

1	2	3
		4
6		5

A. 1 and 3   B. 1 and 5   C. 1 and 6   D. 2 and 4   E. 3 and 5

Question 37:

How many different squares (of either 1, 2, 3 or 4 grid squares in side length) can be made using the grid below?

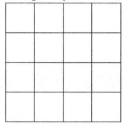

A. 16   B. 20   C. 26   D. 28   E. 30

Question 38:

Dates can be written in a 6 digit format for example 12-10-14 would be 12th October 2014. How many days before this date was the last time when all 6 digits were different?

A. 11   B. 12   C. 13   D. 14   E. 15

Question 39:

A teacher is trying to arrange the 5 students in her class into a seating plan. Ashley must sit on the front row on the left hand side nearest the board because she has poor eyesight. Bella and Caitlin must not be sat in the same row as each other because they talk and disrupt the class. Danielle needs to be sat next to an empty seat as she sometimes has help from a teaching assistant. Emily should be sat on the end of a row because she has poor mobility and it is hard for her to get into a middle seat. The teacher has 2 tables which each sit 3 people, which are arranged 1 behind the other. Who is sitting in the front right seat?

A. Empty seat for the teaching assistant   D. Danielle
B. Bella   E. Emily
C. Caitlin

Question 40:

A packaging company wishes to make cardboard boxes by taking a flat 1.2m by 1.2m square piece of cardboard, cutting square sections out of each corner as shown by the picture below and folding up the sections remaining on each side to make a box. The company experiments with different size boxes by cutting differently sized squares from the corners each time. It makes a box with 10cm by 10cm squares cut out of each corner, a box with 20cm by 20cm squares cut out of each corner and so on up to one with 50cm by 50cm squares cut out of each corner.

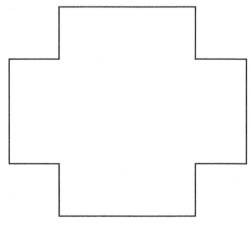

Which side length cut out would result in a box with the largest volume?
A. 10 cm   B. 20 cm   C. 30 cm   D. 40 cm   E. 50 cm

Question 41:

Rental yield for buy-to-let properties is calculated by dividing the potential rent per year paid for a house by the amount it cost to buy the house and get it in a rentable condition. Tina is considering 5 houses as possible buy-to-let investments. House A is in good condition and could be rented as it is for £700 a month, and costs £168,000 to buy. House B is also in good condition but is a student house so Tina would need to buy furniture for it. The house would cost £190,000 to buy and £10,000 to furnish, but could be rented for 40 weeks of the year to 4 students at a rent of £125 a week each. House C needs a lot of work doing. It costs £100,000 but would need £44,000 of renovations, and would rent for £600 a month. House D costs £200,000 and would need £40,000 of renovations, and would rent out for £2000 a month. House E costs £80,000 and would need £20,000 of renovations, and could be rented out for £200 a week.

Which house has the highest rental yield?
A. A   B. B   C. C   D. D   E. E

Question 42:

Wendy is sending 50 invitations to her housewarming party by first class post. Every envelope contains an invitation weighing 70g, and some who are going to family and friends who live further away also contain a sheet of directions, which weigh 25g. The table below gives the prices of sending letters of certain weights by first or second class post. If the total cost of sending the invitations is £33, how many of the invitations contain the extra information?

	First Class	Second Class
Less than 50g	£0.50	£0.30
Less than 75g	£0.60	£0.40
Less than 100g	£0.70	£0.50
Less than 125g	£0.80	£0.60
Less than 150g	£0.90	£0.70

A. 15   B. 20   C. 25   D. 30   E. 35

Question 43:

If the mean of 5 numbers is 8, the median is 6 and the mode is 4, what must the two largest numbers in the set of numbers add up to?
A. 13   B. 16   C. 22   D. 26   E. 28

Question 44:
"St John's Hospital in Northumbria is looking to recruit a new consultant cardiologist, and interviews a series of candidates. The interview panel determines that 3 candidates are clearly more qualified for the role than the others, and they invite these 3 candidates for a second interview. During this second interview, and upon further examination of their previous employment records, it becomes apparent that Candidate 3 is the most proficient at surgery of the 3 candidates, whilst Candidate 1 is the best at patient interaction and explaining the risks of procedures. Candidate 2, meanwhile, ranks between the other 2 in both these aspects.
The hospital director tells the interviewing team that the hospital already has a well-renowned team dedicated to patient interaction, but the surgical success record at the hospital is in need of improvement. The director issues instructions that therefore, it is more important that the new candidate is proficient at surgery, and patient interaction is less of a concern."

Which of the following is a conclusion that can be drawn from the directors' comments?
A. The interviewing team should hire Candidate 2, in order to achieve a balance of good patient relations with good surgical records.
B. The interviewing team should hire Candidate 1, in order to ensure good patient interactions, as these are a vital part of a doctor's work.
C. The interviewing team should ignore the hospital director and assess the candidates further to see who would be the best fit.
D. The interviewing team should hire Candidate 3, in order to ensure that the new candidate has excellent surgical skills, to boost the hospital's success in this area.

Question 45:
A child's fare on the train is more than half but less than three quarters of the adult fare. Both adult and child fares are multiples of 20p. The cost for an adult and 2 children on the train is £4.20. How much is the difference between the adult fare and the child fare?
A. 60p   B. 80p   C. £1   D. £1.20   E. £1.40

Question 46:
A children's football tournament involves a group stage, then a knockout stage. In the group stage, 5 teams play in a round robin format (i.e. each team plays each other once) and the 2 who win the most matches proceed through to a knockout stage where 2 teams play each other and the one that wins proceeds to the next round until there are 2 teams left, who play the final. If we start with 40 teams, how many matches are played altogether?
A. 95   B. 96   C. 97   D. 98   E. 99

Question 47:
I write my 4 digit pin number down in a coded format, by multiplying the first and second number together, dividing by the third number than subtracting the fourth number. If my code is 3, which of these could my pin number be?
A. 3461   B. 9864   C. 5423   D. 7848   E. 6849

Question 48:
Niall can choose whether to join the gym or pay per session. Gym membership costs £30 a month and to attend classes or gym sessions as a member is free. Pay as you go gym sessions cost £4 and attending classes costs £2 each. Niall works out that it will cost him £2 more to pay per session than it will to buy membership. Which of these is a possible combination of Niall's gym sessions and classes for the month?
A. 5 gym sessions, 4 classes   D. 4 gym sessions, 6 classes
B. 4 gym sessions, 4 classes   E. 5 gym sessions, 8 classes
C. 5 gym sessions, 6 classes

Question 49:

Health and safety law dictates that I can only lift boxes of books that weigh in total up to half my weight. I have to move 120 books weighing 2kg each in boxes that when empty weigh 0.5kg. If I weigh 60kg, how many trips will I have to make to move all the books?

A. 5   B. 6   C. 7   D. 8   E. 9

Question 50:

Jenny joined her company 5 and a half years ago on a salary of £24000. At the end of every year of service her pay is reviewed and she gets a pay rise of 3y percent where y is the number of years she has been at the company, rounded up to the nearest £1000.

What is her salary now?

A. 36000   B. 37000   C. 38000   D. 39000   E. 40000

END OF SECTION

YOU MUST ANSWER <u>ONLY</u> <u>ONE</u> OF THE FOLLOWING QUESTIONS

Question 1:
Describe a scientific model of a human brain. Justify any materials you might need, how you would connect the separate components and highlight any limitations your model might have.

Question 2:
"Impatience is the driving force behind development."

To what extent is patience no longer a virtue in modern society?

Question 3:
Voltaire once said, *"The perfect is the enemy of the good."*

Discuss this claim.

Question 4:
Is world peace achievable?

END OF SECTION

Mock Paper E

Question 1:

A square sheet of paper is 20cms long. How many times must it be folded in half before it covers an area of 12.5cm^2?

A. 3   B. 4   C. 5   D. 6   E. 7

Question 2:

Mountain climbing is viewed by some as an extreme sport, while for others it is simply an exhilarating pastime that offers the ultimate challenge of strength, endurance, and sacrifice. It can be highly dangerous, even fatal, especially when the climber is out of his or her depth, or simply gets overwhelmed by weather, terrain, ice, or other dangers of the mountain. Inexperience, poor planning, and inadequate equipment can all contribute to injury or death, so knowing what to do right matters.

Despite all the negatives, when done right, mountain climbing is an exciting, exhilarating, and rewarding experience. This article is an overview beginner's guide and outlines the initial basics to learn. Each step is deserving of an article in its own right, and entire tomes have been written on climbing mountains, so you're advised to spend a good deal of your beginner's learning immersed in reading widely. This basic overview will give you an idea of what is involved in a climb.

Which statement best summarises this paragraph?
A.   Mountain climbing is an extreme sport fraught with dangers.
B.   Without extensive experience embarking on a mountain climb is fatal.
C.   A comprehensive literature search is the key to enjoying mountain climbing.
D.   Mountain climbing is difficult and is a skill that matures with age if pursued.
E.   The terrain is the biggest unknown when climbing a mountain and therefore presents the biggest danger.

Question 3:

50% of an isolated population contract a new strain of resistant Malaria. Only 20% are symptomatic of which 10% are female. What percentage of the total population do symptomatic males represent?

A) 1%   B) 9%   C) 10%   D) 80%

Question 4:

John is a UK citizen yet is looking to buy a holiday home in the South of France. He is purchasing his new home through an agency. Unlike a normal estate agent, they offer monthly discount sales of up to 30%. As a French company, the agency sells in Euros. John decides to hold off on his purchase until the sale in the interest of saving money. What is the major assumption made in doing this?

A.   The house he likes will not be bought in the meantime.
B.   The agency will not be declared bankrupt.
C.   The value of the pound will fall more than 30%.
D.   The value of the pound will fall less than 30%.
E.   The value of the euro may increase by up to 35% in the coming weeks.

Question 5:

In childcare professions, by law, there must be an adult to child ratio of no more than 1:4. Child minders are hired on a salary of £8.50 an hour. What is the maximum number of children that can be continually supervised for a period of 24 hours on a budget of £1,000?

A. 1   B. 8   C. 12   D. 16   E. 468

Question 6:

A table of admission prices for the local cinema is shown below:

	Peak	Off-peak
Adult	£11	£9.50
Child	£7	£5.50
Concession	£7	£5.50
Student	£5	£5

How much would a group of 3 adults, 5 children, a concession and 4 students save by visiting at an off-peak time rather than a peak time?

A. £11.50   B. £13.50   C. £15.50   D. £17.50   E. £18.50

Question 7:

All musicians play instruments. All oboe players are musicians. Oboes and pianos are instruments. Karen is a musician. Which statement is true?

A) Karen plays two instruments.
B) All musicians are oboe players.
C) All instruments are pianos or oboes.
D) Karen is an oboe player.
E) None of the above.

Question 8:

Flow mediated dilatation is a method used to assess vascular function within the body. It essentially adopts the use of an ultrasound scan to measure the percentage increase in the width of an artery before and after occlusion with a blood pressure cuff. Ultrasound scans are taken by one sonographer, and the average lumen diameter is then measured by an analyst. What is a potential flaw in the methodology of this technique?

A) Results will not be comparable within an individual if different arteries start at different diameters.
B) Results will not be comparable between individuals if they have different baseline arterial diameters.
C) Ultrasound is an outdated technique with no use in modern medicine.
D) This methodology is subject to human error.
E) This methodology is not repeatable.

Question 9:

If it takes 20 minutes to board an aeroplane, 15 minutes to disembark and the flight lasts two and a half hours. In the event of a delay it is not uncommon to add 20 minutes to the flight time. Megan is catching the flight in question as she needs to attend a meeting at 5pm. The location of the meeting is 15 minutes from the airport without traffic; 25 minutes with. Which of the following statements is valid considering this information?

A. If Megan wants to be on time for her meeting, given all possibilities described, the latest she can begin boarding at the departure airport is 1.30pm.
B. If Megan starts boarding at 1.40pm she will certainly be late.
C. If Megan aims to start boarding at 1.10pm she will arrive in time whether the plane is delayed or not.
D. If Megan wishes to be on time she doesn't have to worry about the plane being delayed as she can make up the time during the transport time from the arrival airport to the meeting.

Question 10:

A cask of whiskey holds a total volume of 500L. Every two and a half minutes half of the total volume is collected and discarded. How many minutes will it take for the entire cask to be emptied?

A) 80　　　　　B) 160　　　　　C) 200　　　　　D) 240　　　　　E) ∞

Question 11:

B is right of A. C is left of B. D is in front of C. E is in front of B. Where is D in relation to E?

A) D is behind E.
B) E is behind D.
C) D is to the right of E.
D) D is to the left of E.
E) E is to the left of D.

Question 12:

Car A has a fuel tank capacity of 30 gallons and achieves 40mpg. Car B on the other hand has a fuel tank capacity of 50 gallons but only achieves 30mpg. Both cars drive until they run out of fuel. If car A starts with a full tank of petrol and travels 200 miles further than car B, how full was car B's fuel tank?

A) 1/5　　　　　B) 1/4　　　　　C) 1/3　　　　　D) 1/2　　　　　E) 2/3

The information below relates to questions 13 and 14:

The art of change ringing adopts the use of 6 bells, numbered 1 to 6 in order of weight (1 being the lightest). Initially the bells are rung in this order: 1, 2, 3, 4, 5, 6 however the aim is to ultimately ring all the possible combinations of a 6-number sequence. The rules for doing this are very simple: each bell can only move a maximum of one place in the sequence every time it rings.

Question 13:
What is the total possible number of permutations of 6 bells?
A) 160　　　　　B) 220　　　　　C) 660　　　　　D) 720　　　　　E) 1160

Question 14:
Based on the information provided which of the following could be a possible series of bell sequences?

A.	B.	C.	D.	E.
1 2 3 4 5	1 2 3 4 5	1 2 3 4 5	1 2 3 4 5	1 2 3 4 5
2 1 4 3 5	2 1 4 3 5	4 2 1 3 5	1 4 3 2 5	4 1 3 2 5
2 3 1 5 4	2 4 1 5 3	4 2 3 1 5	1 2 3 4 5	5 3 1 2 4

Question 15:
The keypad to a safe comprises the digits 1 - 9. The code itself can be of indeterminate length. The code is therefore set by choosing a reference number so that when a code is entered the average of all the numbers entered must equal the chosen reference number.

Which of the following is true?

A)   If the reference number was set greater than 9, the safe would be locked forever.
B)   This safe is extremely insecure as if random digits were pressed for long enough it would average out at the correct reference number.
C)   More than one number is always required to achieve the reference number.
D)   All of the above are true.
E)   None of the above are true.

Question 16:
The use of antibiotics is one of the major paradoxes in modern medicine. Antibiotics themselves provide a selection pressure to drive the evolution of antibiotic resistant strains of bacteria. This is largely due to the rapid growth rate of bacterial colonies and asexual cell division. As such a widespread initiative is in place to limit the prescription of antibiotics.
Which of the following is a fair assumption?

A)   Antibiotic resistance is impossible to avoid as it is driven by evolution.
B)   If bacteria reproduced at a slower rate antibiotic resistance would not be such an issue.
C)   Medicine always creates more problems than it solves.
D)   In the past antibiotics were used frivolously.
E)   All of the above could be possible.

The information below relates to questions 17 – 21:
The Spaghetti Bolognese recipe below serves 10 people and each portion contains 300 kcal.

- 1kg mince
- 220g pancetta, diced
- 30g crushed garlic
- 1kg tinned tomatoes
- 300g diced onions
- 300g sliced mushrooms
- 200g grated cheese

Question 17:
What quantity of cheese is required to prepare a meal for 350 people?

A) 0.7kg   B) 7kg   C) 70kg   D) 700kg   E) 7000kg

Question 18:
If 12 portions represent 120% of an individual's recommended calorific intake, what is that individuals recommended calorific intake?

A) 2,600kcal   B) 2,800kcal   C) 3,000kcal   D) 3,200kcal   E) 3,400kcal

Question 19:
The recommended ratio of pasta to Bolognese is 4:1. If cooking for 30 people how much pasta should be used?

A) 30.3kg   B) 36.6kg   C) 42.9kg   D) 49.2kg   E) 55.5kg

Question 20:
What is the ratio of onions to the rest of the ingredients if garlic and pancetta are ignored?

A) 1/2.05   B) 1/3.9   C) 1/6.7   D) 1/9.3   E) 1/10

Question 21:

It takes 4 minutes to prepare the ingredients per portion, and a further 8 minutes per portion to cook. Simon has ample preparation space but is limited to cooking 8 portions at a time. What is the shortest period of time it would take him to turn all the ingredients into a meal for 25 people, assuming he didn't start cooking until all the ingredients were prepared?

A)   3 hours   B)   3 hours 40   C)   4 hours   D)   4 hours 40   E)   5 hours

Question 22:

A company sells custom design t-shirts. A breakdown of their costs is shown below:

Number of Items	Cost per Item	
	Black and white	Colour
0 – 99	£3.00	£5.00
100 - 499	£2.50	£4.50
500 - 999	£2.00	£4.00
1000+	£1.00	£3.00

Customers with a never before printed design must also pay a surcharge of £50 to cover the cost of building a jig. What is the total cost for an order of unique stag do t-shirts: 50 in colour, and 200 in black and white?

A)   £650   B)   £700   C)   £750   D)   £800   E)   £850

Question 23:

The Scouts is a movement for young people first established by Lord Baden Powell. As the founder he was the first chief scout of the association. Since his initial appointment there have been a number of notable chief scouts including Peter Duncan and Bear Grylls. Some of the first camping trips conducted by Lord Powell's scout troop were on Brown Sea Island.

Now the Scout movement is a worldwide global phenomenon giving children from all backgrounds the opportunity not only to embark upon adventure but also to engage in the understanding and teaching of foreign culture. Traditionally religion formed the back bone of the scouting movement which was reflected in the scouts promise: "I promise to do my duty to god and to the queen".

Which of the following applies to the scout movement?

A)   Scouts work for the Queen.
B)   The scout network is aimed at adventurous individuals.
C)   Chief scout is appointed by the Queen.
D)   You have to be religious to be a scout.
E)   None of the above.

Question 24:

Three rats are placed in a maze that is in the shape of an equilateral triangle. They pick a direction at random and walk along the side of a triangle. Sophie thinks they are less likely to collide than not. Is she correct?

A)   Yes, because mice naturally keep away from each other.
B)   No, they are more likely to collide than not.
C)   No, they are equally likely to collide than not collide.
D)   Yes, because the probability they collide is 0.25.
E)   None of the above.

Question 25:
The use of human cadavers in the teaching of anatomy is hotly debated. Whilst many argue that it is an invaluable teaching resource, demonstrating far more than a text book can. Others describe how it is an outdated method which puts unfair stress on an already bereaved family. One of the biggest pros for using human tissue in anatomical teaching is the variation that it displays. Whilst textbooks demonstrate a standard model averaged over many 100s of specimens, many argue that it is the variation between cadavers that really reinforces anatomical knowledge.

The opposition argues that it is a cruel process that damages the grieving process of the effected family. For the use of the cadaver often occupies a period of up to 12 months. As such the relative in question is returned to the bereaved family for burial around the time it would be expected that they were recovering as described in the grieving model.

Does the article support or reject the use of cadavers in anatomical teaching?

A) Supports the use   C) Impartial   E) None of the above
B) Rejects the use   D) Can't tell

Question 26:
A ferry is carrying its full capacity. At the time of departure (7am) the travel time to the nearest hour is announced as 13 hours. What is the latest that the ferry could arrive at its destination?

A) 08.00   B) 20.00   C) 20.29   D) 20.30   E) 20.50

Question 27:
A game is played using a circle of 55 stepping stones. A die is rolled showing the numbers 1 - 6. The number on the die tells you how many steps you may take during your go. The only rule is that during your go you must take your steps in the routine two steps forward, 1 step back.

What is the minimum number or rolls required to win?

A) 28   B) 55   C) 110   D) 165   E) 200

Question 28:
On a race track there are 3 cars recording average lap times of 40 seconds, 60 seconds, and 70 seconds. They all started simultaneously 4 minutes ago. How much longer will the race need to continue for them to all cross the start line again at the same time?

A. 23.33 hours   B. 46.67 hours   C. 60.00 hours   D. 83.33 hours   E. 106.67 hours

Question 29:
A class of 60 2nd year medical students are conducting an experiment to measure the velocity of nerve conduction along their radial arteries. This work builds on a previous result obtained demonstrating the effects of how right-handed men have faster nerve conduction velocities than gender matched left handed individuals. 60% of the class are female of which 3% were unable to take part due to underlying heart conditions. 2 of the male members of the class were also unable to take part. On average the female cohort had faster nerve conduction velocities than men in their dominant arm.

Right handed women have the fastest nerve conduction velocities.

A) True   B) False   C) Can't tell

Question 30:

Mark is making a double tetrahedron dice by joining two square based pyramids together at their bases. Each square based pyramid is 5cm wide and 8cm tall. What area of card would have been required to produce the nets for the whole die?

A) $150cm^2$ B) $180 \ cm^2$ C) $210 \ cm^2$ D) $240 \ cm^2$ E) $270 \ cm^2$

Question 31:

A serial dilution is performed by lining up 10 wells and filling each one with 9ml of distilled water. 1 ml of a concentrated solvent is then added to the first well and mixed. 1 ml of this new solution is drawn from the first well and added to the second and mixed. The process is repeated until all 10 wells have been used.

If the solvent starts off at concentration x, what will its final concentration be after 10 wells of serial dilution?

A. $x/10^9$ B. $x/10^{10}$ C. $x/10^{11}$ D. $x/10^{12}$ E. $x/10^{13}$

Question 32:

A student decides to measure the volume of all the blood in his body. He does this by injecting a known quantity of substrate into his arm, waiting a period of 20 minutes, then drawing a blood sample and measuring the concentration of the substrate in his blood.

What assumption has he made here?

A)   The substrate is only soluble in blood.
B)   The substrate is not bioavailable.
C)   The substrate is not excreted.
D)   The substrate is not degraded.
E)   All of the above.

Question 33:

Jason is ordering a buffet for a party. The buffet company can provide a basic spread at £10 per head. However more luxurious items carry a surcharge. Jason is particularly interested in cup-cakes and shell fish. With these items included the buffet company provides a new quote of £10 per head. In addition to simply ordering the food Jason must also purchase cutlery and plates. Plates come in packs of 20 for £8 whilst cutlery is sold in bundles of 60 sets for £10.

With a budget of £2,300 (to the nearest 10 people) what is the maximum number of people Jason can provide food on a plate for?

A)   180   B)   190   C)   209   D)   210   E)   220

Question 34:

What were once methods of hunting have now become popular sports. Examples include archery, the javelin throw, the discuss throw and even throwing a boomerang. Why such dangerous hobbies have begun to thrive is now being investigated by social scientists. One such explanation is that it is because they are dangerous we find them appealing in the first place. Others argue that it is a throwback to our ancestral heritage, where as a hunter gatherer being a proficient hunter was something to show off and flaunt. Whilst this may be the case it is well observed that many find the chase of a hunt exciting if not controversial.

Sports like archery provide excitement analogous to that of the chase during a hunter gatherer hunt.

A)   True
B)   False
C)   Can't tell

Question 35:

"If vaccinations are now compulsory because society has decided that they should be forced, then society should pay for them." Which of the following statements would weaken the argument?

A) Many people disagree that vaccinations should be compulsory.
B) The cost of vaccinations is too high to be funded locally.
C) Vaccinations are supported by many local communities and GPs.
D) Healthcare workers do not want vaccinations.
E) None of the above

Question 36:

Josh is painting the outside walls of his house. The paint he has chosen is sold only in 10L tins. Each tin costs £4.99. Assuming a litre of paint covers an area of $5m^2$, and the total surface area of Josh's outside walls is $1050m^2$; what is the total cost of the paint required if Josh wants to apply 3 coats?

A)   £104.79   B)   £209.58   C)   £314.37   D)   £419.16   E)   £523.95

Question 37:

The stars of the night sky have remained unchanged for many hundreds of years, which allows sailors to navigate using the North Star still to this day. However, this only applies within the northern hemisphere as the populations of the southern hemisphere are subject to an alternative night sky.

An asterism can be used to locate the North Star, it comes by many names including the plough, the saucepan, and the big dipper. Whilst the North Star's position remains fixed in the sky (allowing it to point north reliably always) the rest of the stars traverse around the North Star in a singular motion. In a very long time, the North Star will one day move from its location due to the movement of the Earth.

Which of the following is **NOT** an assumption made in this argument?

A) The Earth is rotating on its axis.
B) Sailors still have need to navigate using the stars.
C) An analogous southern star is used to navigate in the Southern hemisphere.
D) The plough is not the only method of locating the North Star.
E) None of the above.

Question 38:

John wishes to deposit a cheque. The bank's opening times are 9am until 5pm Monday to Friday, 10am until 4pm on Saturdays, and the bank is closed on Sundays. It takes on average 42 bank hours for the money from a cheque to become available.

If John needs the money by 8pm Tuesday, what is the latest he can cash the cheque?

A) 5pm the Saturday before
B) 5pm the Friday before
C) 1pm the Thursday before
D) 1pm the Wednesday before
E) 9am the Tuesday before

Question 39:

How many different diamonds are there in the image shown to the right?

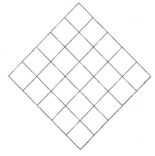

A) 25
B) 32
C) 48
D) 58
E) 63

Question 40:

In 4 years time I will be one third the age that my brother will be next year. In 20 years time he will be double my age. How old am I?

A) 4   B) 9   C) 15   D) 17   E) 23

Question 41:

Aneurysmal disease has been proven to induce systemic inflammatory effects, reaching far beyond the site of the aneurysm. The inflammatory mediator responsible for these processes remains unknown, however the effects of systemic inflammation have been well categorised and observed experimentally in pig models.

This inflammation induces an aberration of endothelial function within the inner most layer of blood vessel walls. The endothelium not only represents the lining of blood vessels but also acts as a transducer converting the haemodynamic forces of blood into a biological response. An example of this is the NO pathway, which uses the shear stress induced by increased blood flow to drive the formation of NO. NO diffuses from the endothelium into the smooth muscle surrounding blood vessels to promote vasodilatation and therefore acts to reduce blood flow.

Failure of this process induces high risk of vascular damage and therefore cardiovascular diseases such as thrombosis and atherosclerosis.

What is a valid implication from the text above?

A) Aneurysmal disease does not affect the NO pathway.
B) Aneurysms directly increase the likelihood of cardiovascular disease.
C) Aneurysms are the opposite of transducers.
D) Observations of this kind should be made in humans to see if the results can be replicated.
E) Aneurysms induce high blood flow.

Question 42:

A traffic surveyor is stood at a T-junction between a main road and a side street. He is only interested in traffic leaving the side street. He logs the class of vehicle, the colour and the direction of travel once on the main road. During an 8-hour period he observes a total of 346 vehicles including bikes. Of which 200 were travelling west whilst the rest travelled east. The over whelming majority of vehicles seen were cars at 90% with bikes, vans and articulated lorries together comprising the remaining 10%. Red was the most common colour observed whilst green was the least. Black and white vehicles were seen in equal quantities.

Which of the following is an accurate inference based on his survey?

A) Global sales are highest for those vehicles which are coloured red.
B) Cars are the most popular vehicle on all roads.
C) Green vehicle sales are down in the area that the surveyor was based.
D) The daily average rate of traffic out of a T junction in Britain is 346 vehicles over 8 hours.
E) To the east of the junction is a dead end.

Question 43:

William, Xavier, and Yolanda race in a 100m race. All of them run at a constant speed during the race. William beats Xavier by 20m. Xavier beats Yolanda by 20m. How many metres does William beat Yolanda?

A) 30m   B) 36m   C) 40m   D) 60m   E) 64m

Question 44:
A television is delivered in a box that has volume 60% larger than that of the television. The television is 150cm x 100cm x 10m. How much surplus volume is there?

A) 0.09m^2 B) 0.9 m^2 C) 9 m^2 D) 90 m^2 E) 900 m^2

Question 45:
Matthew and David are deciding where they would like to go camping Friday to Sunday. Upon completing their research, they discover the following:

➤ Whitmore Bay charges £5.50 a night and does not require a booking. The site provides showers, washing up facilities and easy access to a beach

➤ Port Eynon charges £5 a night and a booking is compulsory. However, the site does not provide showers but does have 240V sockets free of charge

➤ Jackson Bay charges £7 a night and is billed as a luxury site with compulsory booking, private showers, toilets, mobile phone charging facilities and kitchens.

David presents the following suggestion:
As Port Eynon is the farthest distance to travel the benefit of its cheap nightly rate is negated by the cost of petrol. Instead he recommends they visit Jackson Bay as it is the shortest distance to travel and will therefore be the cheapest.

Which of the following best illustrates a flaw in this argument?

A) Whitmore bay may be only a few miles further which means the total cost would be less than visiting Jackson Bay.
B) With kitchen facilities available they will be tempted to buy more food increasing the cost.
C) The campsite may be fully booked.
D) There may be a booking fee driving the cost up above that of the other campsites.
E) All of the above.

Question 46:

The manufacture of any new pharmaceutical is not permitted without scrupulous testing and analysis. This has led to the widespread, and controversial use of animal models in science. Whilst it is possible to test cyto-toxicity on simple cell cultures, to truly predict the effect of a drug within a physiological system it must be trialled in a whole organism. With animals cheap to maintain, readily available, rapidly reproducing and not subject to the same strict ethical laws they have become an invaluable component of modern scientific practice.

Which of the following best illustrates the main conclusion of this argument?

A) New pharmaceuticals cannot be approved without animal experimentation.
B) Cell culture experiments are unhelpful.
C) Modern medicine would not have achieved its current standard without animal experimentation.
D) Logistically animals are easier to keep than humans for mandatory experiments.
E) All of the above.

The information below relates to questions 47 – 50:

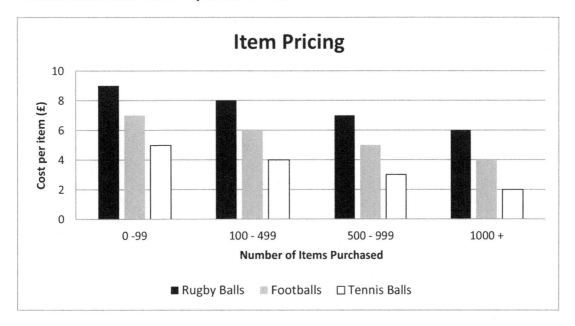

The graph above shows item pricing from a wholesaler. The wholesaler is happy to deliver for a cost of £35 to companies or £5 to individuals. Any order over the cost of £100 qualifies for free delivery. Items are defined as how they come to the wholesaler therefore 1 item = 2 rugby balls or 1 football or 5 tennis balls.

Question 47:

What is the total cost to an individual purchasing 12 rugby balls and 120 tennis balls?

A) £174   B) £179   C) £208   D) £534   E) £588

Question 48:

A private gym wishes to purchase 10 of everything, how short are they of the free delivery boundary?

A) £5.00   C) £10.00   E) They are already over the
B) £5.01   D) £10.01   minimum

Question 49:

What is the most number of balls that can be bought by an individual with £1,000 pounds.

A. 200   C. 500   E. 1,250
B. 250   D. 1,000

Question 50:

The wholesaler sells all his products for a profit of 120%. If he sells £1,320 worth of goods at his prices, what did he spend on acquiring them himself?

A) £400   C) £800   E) £1,120
B) £600   D) £1,100

END OF SECTION

YOU MUST ANSWER <u>ONLY</u> <u>ONE</u> OF THE FOLLOWING QUESTIONS

1) *'Nature over nurture'*. To what extent do you agree with this statement?

2) Do you think positive discrimination should be implemented in our university admission process?

3) *'Developed countries have a responsibility to help developing countries grow'.*

   Do you agree with this statement?

4) Should our education system place a greater emphasis on the sciences as opposed to the arts?

END OF TEST

Mock Paper F

Question 1:

There are four houses on a street. Lucy, Vicky, and Shannon live in adjacent houses. Shannon has a black dog named Chrissie, Lucy has a white Persian cat and Vicky has a red parrot that shouts obscenities. The owner of a four-legged pet has a blue door. Vicky has a neighbour with a red door. Either a cat or bird owner has a white door. Lucy lives opposite a green door. Vicky and Shannon are not neighbours. What colour is Lucy's door?

A) Green

B) Red

C) White

D) Blue

E) Cannot tell

Question 2:

A train driver runs a service between Cardiff and Merthyr. On average a one-way trip takes 40 minutes to drive but he requires 5 minutes to unload passengers and a further 5 minutes to pick up new ones. As the crow flies the distance between Cardiff and Merthyr is 22 miles.

Assuming he works an 8-hour shift with two 20-minute breaks, and when he arrives to work the first train is already loaded with passengers how far does he travel?

A) 132

B) 143

C) 154

D) 176

E) 198

Question 3:

The massive volume of traffic that travels down the M4 corridor regularly leads to congestion at times of commute morning and evening. A case is being made by local councils in congestion areas to introduce relief lanes thus widening the motorway in an attempt to relieve the congestion. This would involve introducing either a new 2 or 4 lanes to the motorway on average costing 1 million pound per lane per 10 miles.

Many conservationist groups are concerned as this will involve the destruction of large areas of countryside either side of the motorway. They argue that the side of a motorway is a unique habitat with many rare species residing there.

The local councils argue that with many hundreds if not thousands of cars siding idle on the motorway pumping pollutants out into the surrounding areas, it is better for the wildlife if the congestion is eased and traffic can flow through. The councils have also remarked that if congestion is eased there would be less money needed to repair the roads from car incidents with could in theory be given to the conservationist groups as a grant.

Which of the following is assumed in this passage?

A) Wildlife living on the side of the motorway cannot be re-homed.

B) Congestion causes car incidents.

C) Relief lanes have been proven to improve traffic jams.

D) A and B.

E) B and C.

F) All of the above.

G) None of the above.

Question 4:

Apples and oranges are sold in packs of 5 for the price of £1 and £1.25 respectively. Alternatively, apples can be purchased individually for 30p and oranges can be purchased individually for 50p. Helen is making a fruit salad, she remarks that her order would have cost her an extra £6.25 if she had purchased the fruit individually.

Which of the following could have been her order?

A) 15 apples 10 oranges

B) 15 apples 15 oranges

C) 25 apples 10 oranges

D) 25 apples, 15 oranges

E) 30 apples, 30 oranges

Question 5:
Janet is conducting an experiment to assess the sensitivity of a bacterial culture to a range of antibiotics. She grows the bacteria so they cover an entire petri dish and then pipettes a single drop of differing antibiotic at different locations. A schematic of her results is shown right where black represents growth of bacteria.

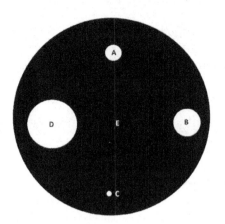

Which of the following best describes Janet's results?

A) This strain of bacteria is susceptible to all antibiotics used.
B) This strain of bacteria is susceptible to none of the antibiotics used.
C) E was the most effective antibiotic.
D) C was the most ineffective.
E) D is the most likely to be used in further testing.

Question 6:
Laura is blowing up balloons for a birthday party. The average volume of a balloon is 300cm³ and Laura's maximum forced expiratory rate in a single breathe is 4.5l/min.

What is the fastest Laura could inflate 25 balloons assuming it takes her 0.5 seconds to breathe in per balloon in and somebody else ties the balloons for her?
A) 112.5 seconds   C) 132.5 seconds   E) 152.5 seconds
B) 122.5 seconds   D) 142.5 seconds

Question 7:
George reasons that A is equal to B which is not equal to C. In which case C is equal to D which is equal to E. Which of the following, if true, would most *weaken* George's argument?

A) A does not equal D.   C) A and B are not equal.   E) None of the above
B) B is equal to E.   D) C is equal to 0.

Question 8:
In a single day how many times do the hour, minute and second hands of analogue clock all point to the same number?

A) 12   B) 24   C) 36   D) 48   E) 72

Question 9:
"People who practice extreme sports should have to buy private health insurance."

Which of the following statements most strongly supports this argument?

A) Exercise is healthy and private insurance offers better reward schemes.
B) Extreme spots have a higher likelihood of injury.
C) Healthcare should be free for all.
D) People that practice extreme sports are more likely to be wealthy.

Question 10:

Explorers in the US in the 18th Century had to contest with a great variety of obstacles ranging from natural to man-made. Natural obstacles included the very nature and set up of the land, presenting explorers with the sheer size of the land mass, the lack of reliable mapping as well as the lack of paths and bridges. On a human level, challenges included the threat from outlaws and other hostile groups. Due to the nature of the settling situation, availability of medical assistance was sketchy and there was a constant threat of diseases and fatal results of injuries.

Which of the following statements is correct with regards to the above text?

A) Medical supply was good in the US in the 18th Century.
B) The land was easy to navigate.
C) There were few outlaws threatening the individual.
D) Crossing rivers could be difficult.
E) All the above.

Question 11:

The statement "The human race is not dependent on electricity" assumes what?

A) We have no other energy resource.
B) Electricity is cheap.
C) Electrical appliances dominate our lives.
D) Electricity is now the accepted energy source and is therefore the only one available.
E) All of the above.

Question 12:

Wine is sold in cases of 6 bottles. A bottle of wine holds 70cl of fluid whereas a wine glass holds 175ml. Cases of wine are currently on offer for £42 a case buy one get one free.

If Elin is hosting a 3-course dinner party for 27 of her friends, and she would like to provide everyone with a glass of wine per course, how much will the wine cost her?

A) £42   B) £84   C) £126   D) £168   E) £210

Question 13:

Hannah buys a television series in boxset. It contains a full 7 series with each series comprising 12 episodes. Rounded to the nearest 10 each episode lasts 40 minutes.

What is the shortest amount of time it could possibly take to watch all the episodes back to back?

A) 49 hours   B) 51 hours   C) 53 hours   D) 56 hours   E) 60 hours

Question 14:

Many are familiar with the story that aided in the discovery of the "germ". Semmelweis worked in a hospital where maternal death rates during labour were astronomically high. He noticed that medical students often went straight from dissection of cadavers to the maternity wards. As an experiment Semmelweis split the student cohort in half. Half did their maternity rotation instead before dissection whereas the other half maintained their traditional routine. In the new routine, maternity ward before dissection, Semmelweis recorded an enormous reduction in maternal deaths and thus the concept of the pathogen was born.

What is best exemplified in this passage?

A) Science is a process of trial and error.
B) Great discoveries come from pattern recognition.
C) Provision of healthcare is closely associated with technological advancements.
D) Experiments always require a control.
E) All of the above.

Question 15:

Jack sits at a table opposite a stranger. The stranger says here I have 3 precious jewels: a diamond, a sapphire, and an emerald. He tells Jack that if he makes a truthful statement Jack will get one of the stones, if he lies he will get nothing.

What must Jack say to ensure he gets the sapphire?

A) Tell the stranger his name.
B) Tell the stranger he must give him the sapphire.
C) Tell the stranger he wants the emerald.
D) Tell the stranger he does not want the emerald or the diamond.
E) Tell the stranger he will not give him the emerald or the diamond.

Question 16:

Simon invests 100 pounds in a saver account that awards compound interest on a 6-monthly basis at 50%. Simon's current account awards compound interest on a yearly basis at 90%.

After 2 years will Simon's investment in the saver account yield more money than it would have in the current account?

A) Yes   B) No   C) Can't tell

Question 17:

My mobile phone has a 4-number pin code using the values 1 – 9. To determine this, I use a standard algorithm of multiplying the first two numbers, subtracting the third and then dividing by the fourth. I change the code by changing the answer to this algorithm – I call this the key. What is the largest possible key?

A) 42
B) 55
C) 70
D) 80
E) 81

Question 18:

A group of scientists investigates the role of different nutrients after exercise. They set up two groups of averagely fit individuals consisting of the same number of both males and females aged 20 – 25 and weighing between 70 and 85 kilos. Each group will conduct the same 1hr exercise routine of resistance training, consisting of various weighted movements. After the workout they will receive a shake with vanilla flavour that has identical consistency and colour in all cases. Group A will receive a shake containing 50 g of protein and 50g of carbohydrates. Group B will receive a shake containing 100 g of protein and 50 g of carbohydrates. All participants have their lean body mass measured before starting the experiment.

Which of the following statements is correct?

A) The experiment compares the response of men and women to endurance training.
B) The experiment is flawed as it does not take into consideration that men and women respond differently to exercise.
C) The experiment does not consider age.
D) The experiment mainly looks at the role of protein after exercise.
E) None of the above.

Question 19:

Adam, Beth and Charlie are going on holiday together. A single room costs £60 per night, a double room costs £105 per night and a four-person room costs £215 per night. It is possible to opt out from the cleaning service and to pay £12 less each night per room.

What is the minimum amount the three friends could pay for their holiday for a three-night stay at the hotel?

A) £122   B) £144   C) £203   D) £423   E) £432

Question 20:

I have two 96ml glasses of squash. The first is comprised of $\frac{1}{6}$ squash and $\frac{5}{6}$ water. The second is comprised of $\frac{1}{4}$ water and $\frac{3}{4}$ squash. I take 48ml from the first glass and add it to glass two. I then take 72ml from glass two and add it to glass one.

How much squash is now in each glass?

A) 16ml squash in glass one and 72ml squash in glass two.
B) 40ml squash in glass one and 32ml squash in glass two.
C) 48ml squash in glass one and 32ml squash in glass two.
D) 48ml squash in glass one and 40ml squash in glass two.
E) 80ml squash in glass one and 40ml squash in glass two.

Question 21:

It may amount to millions of pounds each year of taxpayers' money; however, it is strongly advisable for the HPV vaccination in schools to remain. The vaccine, given to teenage girls, has the potential to significantly reduce cervical cancer deaths and furthermore, the vaccines will decrease the requirement for biopsies and invasive procedures related to the follow-up tests. Extensive clinical trials and continued monitoring suggest that both Gardasil and Cervarix are safe and tolerated well by recipients. Moreover, studies demonstrate that a large majority of teenage girls and their parents are in support of the vaccine.

Which of the following is the conclusion of the above argument?

A) HPV vaccines are safe and well tolerated
B) It is strongly advisable for the HPV vaccination in schools to remain
C) The HPV vaccine amounts to millions of pounds each year of taxpayers' money
D) The vaccine has the potential to significantly reduce cervical cancer deaths
E) Vaccinations are vital to disease prevention across the population

Question 22:

Anna cycles to school, which takes 30 minutes. James takes the bus, which leaves from the same place as Anna, but 6 minutes later and gets to school at the same time as Anna. It takes the bus 12 minutes to get to the post office, which is 3km away. The speed of the bus is $\frac{5}{4}$ the speed of the bike. One day Anna leaves 4 minutes late.

How far does she get before she is overtaken by the bus?

A) 1.5km   B) 2km   C) 3km   D) 4km   E) 6km

Question 23:

The set two maths teacher is trying to work out who needs to be moved up to set one and who to award a certificate at the end of term. The students must fulfil certain criteria:

	Attendance over 95%
Move to set one	**Average test mark over 92**
	Less than 5% homework handed in late
	Absences below 4%
Awarded a Certificate	**Average test mark over 89**
	At least 98% homework handed in on time

	Terry	Alex	Bahara	Lucy	Shiv
Attendance %	97	92	97	100	98
Average test mark %	89	93	94	95	86
Homework handed in on time %	96	92	100	96	98

Who would move up a set and who would receive a certificate?
A) Bahara would move up a set and receive a certificate.
B) Bahara and Lucy would move up a set and Bahara would receive a certificate.
C) Bahara, Terry and Lucy would move up a set and Bahara and Shiv would receive a certificate.
D) Lucy would move up a set and Bahara would receive a certificate.
E) Lucy would move up a set and Bahara and Terry would receive a certificate.

Question 24:

18 years ago, A was 25 years younger than B is now. In 21 years time, A will be 28 years older than B was 14 years ago. How old is A now if A is $\frac{5}{6}$B?

A) 27   B) 28   C) 35   D) 42   E) 46

Question 25:

The time now is 10.45am. I am preparing a meal for 16 guests who will arrive tomorrow for afternoon tea. I want to make 3 scones for each guest, which can be baked in batches of 6. Each batch takes 35 minutes to prepare and 25 minutes to cook in the oven and I can start the next batch while the previous batch is in the oven. I also want to make 2 cupcakes for each guest, which can be baked in batches of 8. It takes 15 minutes to prepare the mixture for each batch and 20 minutes to cook them in the oven. I will also make 3 cucumber sandwiches for each guest. 6 cucumber sandwiches take 5 minutes to prepare.

What will the time be when I finish making all the food for tomorrow?

A) 4:35pm   B) 5.55pm   C) 6:00pm   D) 6:05pm   E) 7:20pm

Question 26:

Pyramid	Base edge (m)	Volume (m³)
1	3	33
2	4	64
3	2	8
4	6	120
5	2	8
6	6	120
7	4	64

What is the difference between the height of the smallest and tallest pyramids?

A) 1m   B) 5m   C) 4m   D) 6m   E) 8m

Question 27:

The wage of Employees at Star Bakery is calculated as: £210 + (Age x 1.2) – 0.8 (100 - % attendance).
Jessica is 35 and her attendance is 96%. Samira is 65 and her attendance is 89%.

What is the difference between their wages?

A) £30.40   B) £60.50   C) £248.80   D) £263.20   E) £279.20

Question 28:

It is important that research universities demonstrate convincing support of teaching. Undergraduates comprise an overwhelming proportion of all students and universities should make an effort to cater to the requirements of the majority of their student body. After all, many of these students may choose to pursue a path involving research and a strong education would provide students with skills equipped towards a career in research.

What is the conclusion of the above argument?
A) Undergraduates comprise an overwhelming proportion of all students.
B) A strong education would provide a strong foundation and skills equipped towards a career in research.
C) Research universities should strongly support teaching.
D) Institutions should provide undergraduates with a high-quality learning experience.
E) Research has a greater impact than teaching and limited universities funds should mainly be invested in research.

Question 29:

American football has reached a level of violence that puts its players at too high a level of risk. It has been suggested that the NFL, the governing body for American football should get rid of the iconic helmets. The hard-plastic helmets all have to meet minimum impact-resistance standards intended to enhance safety, however in reality they gave players a false sense of security that only resulted in harder collisions. Some players now suffer from early onset dementia, mood swings and depression. The proposal to ban helmets for good should be supported. Moreover, it would prevent costly legal settlements involving the NFL and ex-players suffering from head trauma.

What is the conclusion of the above argument?
A) Sports players should not be exposed to unnecessary danger.
B) Helmets give players a false sense of security.
C) Players can suffer from early onset dementia, mood swings and depression.
D) The proposal to ban helmets should be supported.
E) American football is too violent and puts its players at risk.

Question 30:

At the final stop (stop 6), 10 people get off the tube. At the previous stop (stop 5) $\frac{1}{2}$ of passengers got off. At stop 4, $\frac{3}{5}$ of passengers got off. At stop 3, $\frac{1}{3}$ of passengers got off and at stops 1 and 2, $\frac{1}{6}$ of passengers got off.

How many passengers got on at the first stop?

A) 10   B) 36   C) 90   D) 108   E) 3600

Question 31:

Everyone likes English. Some students born in spring like Maths and some like Biology. All students born in winter like Music and some like Art. Of those born in autumn, no one likes Biology, and everyone likes Art.

Which of the following is true?

A) Some students born in spring like both Biology and Maths.
B) Students born in spring, winter, and autumn all like Art.
C) No one born in winter or autumn likes Biology.
D) No one who likes Biology also likes Art.
E) Some students born in winter like 3 subjects.

Question 32:

Until the twentieth century, the whole purpose of art was to create beautiful, flawless works. Artists attained a level of skill and craft that took decades to perfect and could not be mirrored by those who had not taken great pains to master it. The serenity and beauty produced from movements such as impressionism has however culminated in repulsive and horrific displays of rotting carcasses designed to provoke an emotional response rather than admiration. These works cannot be described as beautiful by either the public or art critics. While these works may be engaging on an intellectual or academic level, they no longer constitute art.

Which of the following is an assumption of the above argument?
A) Beauty is a defining property of art.
B) All modern art is ugly.
C) Twenty first century artists do not study for decades.
D) The impressionist movement created beautiful works of art.
E) Some modern art provokes an emotional response.

Question 33:

The cost of sunglasses is reduced over the bank holiday weekend. On Saturday, the price of the sunglasses on Friday is reduced by 10%. On Sunday the price of the sunglasses on Saturday is reduced by 10%. On Monday, the price of the sunglasses on Sunday is reduced by a further 10%. What percentage of the price on Friday is the price of the sunglasses on Monday?

A) 55.12%   B) 59.10%   C) 63.80%   D) 70.34%   E) 72.9%

Question 34:

Putting the digit 7 on the right-hand side of a two-digit number causes the number to increase by 565. What is the value of the two-digit number?

A) 27
B) 52
C) 62
D) 66
E) 627

Question 35:

When folded, which box can be made from the net shown below?

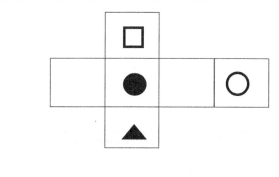

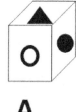   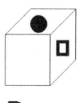 

A   **B**   **C**   **D**   **E**

Question 36:

The grid is comprised of 49 squares. The shape's area is 588cm². What is its perimeter in cm?

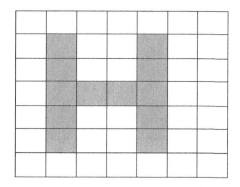

A) 26   B) 49   C) 84   D) 126   E) 182

Questions 37-39 refer to the following information:

$$BMI = weight\ (kg) \div height^2\ (m^2)$$

Men	BMR= (10 x weight in kg) + (6.25 x height in cm) – (5 x age in years) + 5
Women	BMR= (10 x weight in kg) + (6.25 x height in cm) – (5 x age in years) -161

Recommended Intake:

Amount of Exercise	Daily Kilocalories required
Little to no exercise	BMR x 1.2
Light exercise 1-3 days per week	BMR x 1.375
Moderate exercise 3-5 days per week	BMR x 1.55
Heavy exercise 6-7 days per week	BMR x 1.725
Very heavy exercise twice per day	BMR x 1.9

Question 37:
A child weighs 35kg and is 120cm tall. What is the BMI of the child to the nearest 2 decimal places?

A) 0.0024   B) 24.28   C) 24.31   D) 42.01   E) 42.33

Question 38:
What is the BMR of a 32-year-old woman weighing 80kg and measuring 1.7m in height?

A) 643.7 kcal   B) 1537 kcal   C) 1541.5 kcal   D) 1707.5 kcal   E) 2707.5 kcal

Question 39:
What is the recommended intake of a 45-year-old man weighing 80kg and measuring 1.7m in height who does little to no exercise each week?

A) 1642.5 kcal   B) 1771.8 kcal   C) 1851 kcal   D) 1971 kcal   E) 2712.5 kcal

Questions 40 and 41 relate to the following passage:
The achievement levels of teenagers could be higher if school started later. Teenagers are getting too little sleep because they attend schools that start at 8:30am or earlier. A low level of sleep disrupts the body's circadian rhythms and can contribute to health problems such as obesity and depression. Some doctors are now urging schools to start later in order for teenagers to get sufficient sleep; ideally 8.5 to 9.5 hours each night. During adolescence, the hormone melatonin is released comparatively later in the day and the secretion levels climb at night. Consequently, teenagers can have trouble getting to sleep earlier in the night before sufficient melatonin is present. A school in America tested this idea by starting one hour later and the percentage of GCSEs at grades A* - C increased by 16%. Schools in the UK should follow by example and shift start times later.

Question 40:
Which of the following is a flaw in the above argument?

A) Slippery slope.
B) Hasty generalisation.
C) It confuses correlation with cause.
D) Circular argument.
E) Schools should not prioritise academic achievement.

Question 41:

Which of the following, if true, would most strengthen the above argument?

A) Teenagers who are more alert will disrupt the class more.
B) Getting more sleep at night is proven to increase activity levels.
C) American schools and British schools teach the same curriculum.
D) The school in America did not alter any other aspects of the school during the trial for example curriculum, teachers, and number of students in each class.
E) Teachers get tired towards the end of the school day and are less effective.

Question 42:

The UK energy market is highly competitive. In an effort to attract more business and increase revenue, the company EnergyFirst has invested significant funds into its publicity sector. Last month, they doubled their advertising expenditures, becoming the energy company to invest the greatest proportion of investment into advertising. As a result, it is expected that EnergyFirst will expand its customer base at a rate exceeding its competitors in the ensuing months. Other energy companies are likely to follow by example.

Which of the following, if true, is most likely to weaken the above argument?

A) Other companies invest more money into good customer service.
B) Research into the energy industry demonstrates a low correlation between advertising investment and new customers.
C) The UK energy market is not highly competitive.
D) EnergyFirst currently has the smallest customer base.
E) Visual advertising heavily influences customers.

Question 43:

The consumption of large quantities of red meat is suggested to have negative health ramifications. Carnitine is a compound present in red meat and a link has been discovered between carnitine and the development of atherosclerosis, involving the hardening, and narrowing of arteries. Intestinal bacteria convert carnitine to trimethylamine-N-oxide, which has properties that are damaging to the heart. Moreover, red meat consumption has been associated with a reduced life expectancy. It may be that charring meat generates toxins that elevate the chance of developing stomach cancer. If people want to be healthy, a vegetarian diet is preferable to a diet including meat. Vegetarians often have lower cholesterol and blood pressure and a reduced risk of heart disease.

Which of the following is an assumption of the above argument?

A) Diet is essential to health and we should all want to be healthy.
B) Vegetarians do the same amount of exercise as meat eaters.
C) Meat has no health benefits.
D) People who eat red meat die earlier.
E) Red meat is the best source of iron.

Question 44:

Auckland is 11 hours ahead of London. Calgary is 7 hours behind London. Boston is 5 hours behind London. The flight from Auckland to London is 22 hours, but the plane must stop for 2 hours in Hong Kong. The flight from London to Calgary is 8 hours 30 minutes. The flight from Calgary to Boston is 6 hours 30 minutes. Sam leaves Auckland at 10am for London. On arrival to London, he waits 3 hours then gets the plane to Calgary. Once in Calgary, he waits 1.5 hours and gets the plane to Boston. What time is it when Sam arrives in Boston?

A) 13:30pm   B) 22.30pm   C) 01:00am   D) 01:30am   E) 03:30am

Question 45:

Light A flashes every 18 seconds, light B flashes every 33 seconds and light C flashes every 27 seconds. The three lights all flashed at the same time 5 minutes ago.

How long will it be until they next all flash simultaneously?

A) 33 seconds   B) 294 seconds   C) 300 seconds   D) 333 seconds   E) 594 seconds

Question 46:

There are 30 students in a class. They must all play at least 1 instrument, but no more than 3 instruments. 70% play the piano, 40% play the violin, 20% play the guitar and 10% play the saxophone. Which of the following statements must be true?

1. 3 or more students play piano and violin.
2. 12 students or less play the piano and the violin.
3. 9 or more students do not play piano or violin.

A) 1 only   B) 1 and 2   C) 2 only   D) 2 and 3   E) 3 only

Question 47:

Neil, Simon and Lucy are playing a game to see who can role the highest number with two dice. They start with £50 each. The losers must halve their money and give it to the winner of each game. If it is a draw, the two winners share the loser's money. If all three tie, then they keep their money. Neil wins game 1, Simon and Lucy win game 2 and 5 and Lucy wins games 3 and 4.

How much money does Lucy gain?

A) £15.63   B) £75.00   C) £75.16   D) £78.13   E) £128.10

Question 48:

Drivers in the age group 17-19 comprise 1.5% of all drivers; however, 12% of all collisions involve young drivers in this age category. The RAC Foundation wants a graduated licensing system with a 1-year probationary period with restrictions on what new drivers can do on roads. Additionally, driving instructors need to emphasise the dangers of driving too fast and driving tests should be designed to make new drivers more focused on noticing potential hazards. These changes are essential and could stop 4,500 injuries on an annual basis.

What is the assumption of the above argument?
A) Young drivers are more likely to have more passengers than other age groups.
B) Young drivers spend more hours driving than older drivers.
C) Young drivers are responsible for the collisions.
D) The cars that young people drive are unsafe.
E) Most young drivers involved in accidents are male.

Question 49:

The mean weight of 6 apples is 180g. The lightest apple weighs 167g. What is the highest possible weight of the heaviest apple?

A) 193g   B) 225g   C) 235g   D) 245g   E) 255g

Question 50:

"Sugar should be taxed like alcohol and cigarettes." Which of the following arguments most supports this claim?

A) Sugar can cause diabetes.
B) Sugar has high addictive potential and is associated with various health concerns.
C) High sugar diets increase obesity.
D) People that eat a lot of sugar are more likely to start abusing alcohol.
E) None of the above.

END OF SECTION

YOU MUST ANSWER <u>ONLY</u> <u>ONE</u> OF THE FOLLOWING QUESTIONS

1   *'Social media has risen and fallen'.*

Discuss.

2   Do you think the right to press freedom has become increasingly dangerous and polarising to society?

3   *'Artificial intelligence replaces humans and leads to unemployment'.*

Do you agree with this statement?

4   Does our society place too much emphasis on a university education?

END OF TEST

ANSWERS

Answer Key

Paper A		Paper B		Paper C		Paper D		Paper E		Paper F	
1	B	1	C	1	D	1	C	1	C	1	B
2	D	2	B	2	A	2	D	2	D	2	E
3	E	3	B	3	C	3	B	3	B	3	E
4	C	4	B	4	D	4	E	4	D	4	D
5	C	5	D	5	A	5	A	5	D	5	E
6	C	6	B	6	B	6	C	6	B	6	A
7	E	7	D	7	E	7	B	7	A	7	B
8	C	8	C	8	E	8	B	8	D	8	B
9	D	9	B	9	D	9	D	9	C	9	B
10	D	10	C	10	B	10	A	10	E	10	D
11	C	11	E	11	A	11	B	11	D	11	C
12	C	12	A	12	D	12	E	12	E	12	B
13	D	13	C	13	C	13	D	13	D	13	E
14	B	14	B	14	A	14	C	14	E	14	E
15	B	15	C	15	E	15	D	15	A	15	C
16	E	16	E	16	D	16	A	16	D	16	A
17	C	17	D	17	B	17	B	17	B	17	D
18	C	18	E	18	B	18	E	18	C	18	D
19	C	19	A	19	B	19	C	19	B	19	D
20	B	20	D	20	B	20	E	20	D	20	D
21	C	21	E	21	B	21	B	21	E	21	B
22	E	22	E	22	D	22	C	22	D	22	B
23	C	23	D	23	D	23	E	23	E	23	B
24	C	24	B + E	24	D	24	A	24	B	24	C
25	B	25	D	25	B	25	D	25	C	25	D
26	D	26	C	26	A	26	D	26	C	26	D
27	D	27	B	27	D	27	C	27	B	27	A
28	E	28	C	28	B	28	D	28	B	28	C
29	B	29	C	29	A	29	E	29	C	29	D
30	C	30	C	30	E	30	E	30	C	30	D
31	C	31	A	31	A	31	B	31	B	31	E
32	D	32	A	32	C	32	C	32	E	32	A
33	C	33	D	33	B	33	D	33	D	33	E
34	B	34	C	34	C	34	C	34	A	34	C
35	D	35	B	35	E	35	D	35	B	35	E
36	B	36	E	36	E	36	C	36	D	36	E
37	C	37	C	37	D	37	E	37	A	37	E
38	D	38	B	38	A	38	D	38	E	38	C
39	D	39	C	39	B	39	E	39	D	39	D
40	E	40	E	40	C	40	B	40	B	40	B
41	B	41	D	41	A	41	E	41	D	41	D
42	D	42	C	42	E	42	D	42	E	42	B
43	D	43	C	43	A	43	D	43	D	43	A
44	D	44	B	44	D	44	D	44	D	44	A
45	B	45	A	45	E	45	B	45	B	45	B
46	C	46	B	46	C	46	A	46	E	46	B
47	C	47	C	47	B	47	E	47	D	47	D
48	E	48	C	48	B	48	C	48	C	48	C
49	A	49	C	49	E	49	E	49	B	49	D
50	B	50	B	50	D	50	E	50	A	50	B

Raw to Scaled Scores									
1	10.5	**11**	38.5	**21**	49.5	**31**	59.0	**41**	70.5
2	18.0	**12**	40.0	**22**	50.5	**32**	60.0	**42**	72.0
3	22.5	**13**	41.0	**23**	51.5	**33**	61.0	**43**	74.0
4	25.5	**14**	42.0	**24**	52.5	**34**	62.0	**44**	75.5
5	28.5	**15**	43.5	**25**	53.5	**35**	63.0	**45**	78.0
6	30.5	**16**	44.5	**26**	54.5	**36**	64.0	**46**	80.5
7	32.5	**17**	45.5	**27**	55.0	**37**	65.0	**47**	83.5
8	34.0	**18**	46.5	**28**	56.0	**38**	66.5	**48**	88.0
9	35.5	**19**	47.5	**29**	57.0	**39**	67.5	**49**	95.5
10	37.0	**20**	48.5	**30**	58.0	**40**	69.0	**50**	103

Mock Paper Answers

Mock Paper A: Section 1

Question 1: B
James runs 26.2 seconds, which is outside the qualifying time, therefore he does not qualify

Question 2: D
5.6/7 gives the unit price of 80p – this equals a packet of crisps. Multiplying this by 2 gives the sandwich and by 4 gives the watermelon price of £3.20

Question 3: E
Jane leaves at 2:35pm and arrives at 3:25pm, taking 50 minutes. Sam's journey takes twice as long, so leaving at 3:00pm it takes 100 minutes, giving an arrival time of 4:40pm

Question 4: C
After the donation, Michael has eight sweets. Therefore, Hannah had 16 sweets after the transaction and hence 13 sweets before

Question 5: C
Find original pay: £250/0.86 = 290.697674 basic original pay. Add the rise: (290.697674 x 1.05) + 6 = £311.232558 new basic pay. Subtract the income tax at 12% = 311.232558 x 0.88 = £273.884651 new pay rate. To the nearest whole pound this is £274

Question 6: C
Given the first cube is a white cube, you are drawing from one of three boxes, boxes A, C or D. Boxes C and D will have just had their only white cube removed, whereas box A will have one white cube remaining. Therefore, the probability of drawing a second white cube is $^1/_3$, thus the probability of non-white (i.e. black) is $^2/_3$.

Question 7: E
This is a simultaneous equations question. 500 + 10(x – 80) = 600 + 5x; true when x ≥ 80.
500 + 10x – 800 = 600 + 5x
» 5x = 900
» x = 180

Question 8: C

If eating more slowly caused a reduction in the time available to work, the candidate might be less productive.

Question 9: D

This is a LCM question. We need to find the lowest common multiple of the song lengths. The LCM of 100, 180 and 240 is 3,600 seconds – equal to 60 minutes. For ease of arithmetic, you may choose to work reduce all numbers by a factor of 10.

Question 10: D

The journey is 3 hours and 45 mins, minus a 14 minute break gives 3hrs 31 mins travel time, or 211 minutes. Therefore, the average speed is 51mph, or 82kph by using the stated conversion factor.

Question 11: C

The mean guess is £13.80, which is £5.80 too high

Question 12: C

The overall error for respondent 3 is £13, which is the least

Question 13: D

Scale back and forth from known quantities. Country B has 32.1m so Country D has 38.5m people.

Question 14: B

Country B has 32.1m people. Therefore $45 x 32.1m = $1.44bn

Question 15: B

The average speed is 24mph, independent of distance travelled as it cancels. Imagine this covers a set distance of say 30 miles. It will take 1 hour on the way and 1.5 hours on the way back. 60/2.5 = 24. This is true of all distances, the ratio is the same.

Question 16: E

None of the above can be reliably deduced from the passage alone

Question 17: C

Imagine the toothpaste costs 100p originally, and follow the price through. It rises by 80% to 180p, then is reduced by 50% to 90p. Three tubes are purchased for the price of 2 (i.e. 180p), therefore the cost per unit is 180/3 = 60p. 60p = 60% x 100, the original price'

Question 18: C

Argument C has the same form, asserting that since something is not happening, the result of the action will never be true.

Question 19: C

Statement C is the only one making reference to the potential outcome of solving crimes faster, thereby providing a plausible mechanism for a reduction in cybercrime rates

Question 20: B

The passage suggests that the attacks were carried out by extra terrestrial beings. Though the supposed UFO sightings have rational explanations, the writer feels this is insufficient to dismiss his idea.

Question 21: C

The initial argument suggests that two things must be present for an action to happen. If only one is absent, the action cannot happen. Argument C has the same form, the others do not.

Question 22: E
Growing vegetables needs several positive traits. The passage does not tell us which is the most important or most commonly lacked skill, only that more than one skill is required for success.

Question 23: C
Joseph does not have blue cubic blocks, since all his blue block are cylindrical.

Question 24: C
$130°$. Each hour is 1/12 of a complete turn, equalling $30°$. The smaller angle between 4 and 8 on the clock face is 4 gaps, therefore $120°$. In addition, there is 1/3 of the distance between 3 and 4 still to turn, so an additional $10°$ must be added on to account for that.

Question 25: B
The chance of red is 2/6 = 1/3. To get no reds at all, it must be non-red for each of three independent rolls. The probability of this is $(2/3)^3$ = 8/27. Therefore the probability of at least one red is $1 - 8/27 = \underline{19/27}$

Question 26: D
These three furniture items are compatible with having 6 legs. All the other statements are false.

Question 27: D
Work this out by time. The friends are closing on each other at a total of 6mph overall, therefore the 42 miles take 7 hours. In seven hours, the pigeon, flying at 18moh covers 18 x 7 = 126 miles.

Question 28: E
The passage does not make any supported claims about fruit juice. It gives rationale for both benefits and risks of fruit juice consumption without reaching a conclusion.

Question 29: B
Calculate the overall cost of three stationery sets, then subtract any items not bought. For each item shared between two people, there is one of that item not required. The overall cost is £6.00 per person, £18.00 overall. Subtract one geometry set (£3), one paper pad (£1) and one pencil (50p) to give £13.50 overall cost.

Question 30: C
Argument C is the most convincing. It gives a strong rationale as to why the notion that people should only pay for services which they personally use is likely to have serious adverse consequences on the nation as a whole. Therefore this flawed logic is not suitable to apply to the arts funding dilemma.

Question 31: C
Moving matches 1 and 4 to form a cross inside one of the other cubes will solve the problem. Two squares are broken (the top left hand corner and the overall large square) but four new small ones are created, bringing the total up to seven.

Question 32: D
The white square is opposite brown, since both are adjacent to blue on opposite sides. White and brown cannot be adjacent to each other since the position of the opposite black and red sides makes that impossible.

Question 33: C
We take the overall price to the UK and subtract money which does not go to the farmers. 36,000,000kg at 300p/kg gives £108m. Subtract commission 108 x 0.8, then take 10% of the remaining proceeds as the farmers' share, giving £8.64m

Question 34: B
The first paragraph tells us annual road deaths have fallen, so B is true. The others are false.

Question 35: D
Regression to the mean is a phenomenon observed when a value is variable within a probability distribution. Sometimes by chance it will be at the high or low end, but thereafter it is likely to be closer to what is expected. This can explain the fall in drink driving deaths after the new campaign.

Question 36: B
The relevant set is 'people with a sore throat and a chest infection'. If *some* (i.e. not *all*) members of that set have the 'flu', then it follows, necessarily, that the other members of the set do not. This is because you can only either *have* the 'flu', or *not have* the 'flu'. So, B is supported.

Question 37: C
Catherine must choose four socks. If choosing three or fewer, it is possible that they could each be of different colour. When choosing four, it is certain that at least two socks will make a matching pair, but possible that there will be two pairs.

Question 38: D
This is another simultaneous equations question. Solve to find x, the normal rate of pay.
$100x + 20y = 2000$ » $60y = 6000 - 300x$ (substitute this)

$80x + 60y = 2700$
» $80x + (6000 - 300x) = 2700$
» $220x = 3300$
$x = 15$
Substitute x=15 into one of the original equations, and then solve to get $y = 25$

Question 39: D
The easiest way to do this is via simultaneous equations. Let A be the distance travelled by the Plymouth train and B the distance travelled by the Manchester train. Thus:
$A = 90x + 45$ and $B = 70x$
The collision will occur when the total distance travelled by both trains is = 405
i.e. $A + B = 405$
Therefore, $90x + 45 + 70x = 405$
$X = 2.25$ hours. Thus, collision happens at 12:45.
Substitute x=2.25 into the first equation to give the distance from Plymouth:
$A = 90 \times 2.25 + 45 = 247.5$, which rounds to 248 miles

Question 40: E
Statement E is not true, the others are true. A pregnant rabbit requires 50 pieces per day and a normal rabbit requires 25. Therefore three pregnant and ten normal rabbits require only 400 pieces per day, not 450.

Question 41: B
If memory of names uses a different part of the brain, then conclusions drawn from this experiment may have no validity.

Question 42: D
Michael pays £60 and £110 = £170 for the painting. He sells it for £90 and £130 = £220. Thus, he makes a profit of £220 - £170 = £50.

Question 43: D
The principle problem is that it does not compare the relative effectiveness of pesticides and natural predators. It might be that pesticides are far more effective at controlling pests, despite the unnecessary excess killing.

Question 44: D
Proportionately, there would be 172 members. Therefore there is an excess of $298 - 172 = 126$ members.

Question 45: B

If each pair of opposite faces is painted one colour, this requirement can be satisfied with a minimum of three colours.

Question 46: C

Two thin people cross dropping one (first crossing). One returns (second) and then takes another thin person over(third). When one returns this time (fourth) he leaves two on the correct bank and returns for the fat person alone on the wrong side. The fat person comes back alone leaving one thin person on the wrong side (five) and one of the thin people (six) return to collect the final thin person. These two return together completing the 7th crossing.

Question 47: C

Christmas day is a Wednesday. If there are only four Saturdays and four Wednesdays, the 1st December is a Sunday and thus the 25th is a Wednesday.

Question 48: E

Iver is West of Ruddock, but we don't know where it is in relation to Dell as this is not specified, therefore E cannot be concluded. The other statements are true.

Question 49: A

The passage argues that as most women's needs have changed, the style of clothing should change. This is derived from the principle that the needs of the majority should be prioritised.

Question 50: B

This argument has the same structure. It says since one thing is not possible, something more extreme than it with regard to the same characteristic is also not possible. The other arguments have a different construction.

END OF SECTION

Mock Paper A: Section 2

1. *"Strive not to be a success, but to be of value"*
    To what extent is it possible to be "a success", but to have little value?

Introduction:

- Definitions of success and value should be given, with a brief exploration of what is implied by the two terms and any connotations.
- Success could be considered to be defined through achievements or status, whilst value is defined on a more emotional or social level with the influence that people have on the situation around them.
- The introduction should begin to explore where there are overlaps between the two terms and where they could be seen as standing alone. This could be done by general statements or through specific examples, so long as valid ideas are introduced on both sides.

Possible arguments for success without value:

- In the context of money – a person could have a successful career defined by the amount of money that they make, but may not have been a positive influence on those around them. You could consider the extent to which the definitions may get confused with worth but doesn't mean it (e.g. she's worth £xxx as a term for success)
- In the context of academic achievement – just because you have been successful with grades at school, for example, doesn't mean you were necessarily valued in the setting or by those around you.
- In the context of gambling or competitive sports/games – just because a person has been successful in 'winning' doesn't necessarily mean that they were valued as a player or by others around them.

Possible arguments for success and value being integrated:

- Is success a value in itself? Could this be in the eye of the beholder?
- In terms of gaining power or authority – you could choose to explore whether figures in power such as a manager, politician etc. must be valued by somebody to reach this status?
- In particular careers such as teaching or charity work – in order to be successful must you be doing something that is valued by somebody?
- On an emotional level, could you be considered a 'successful' friend or family member if you are valued within this setting? This may be particularly relevant using the example of a mother figure.
- You may wish to put forward the idea that you can be valuable without success, but does this work the other way round?

Conclusion:

- Include a summary of all points given, and refer back to your original definition.
- The conclusion should reach a clear and logical solution – you could choose to completely decide that it is impossible, but most probably will reach a compromise which encompasses the idea that value is distinct from success, but the terms can be used interchangeably in particular situations.

2. *Is the media a positive or negative influence on scientific understanding?*

Introduction:

- Introduce or illustrate the ways in which the media are involved in our scientific understanding – you could describe what ways we might have access to the information, such as newspapers, TV.
- You may give reference to the particular areas of scientific interest in the media, such as health and global warming.
- Introduce both sides of the argument – the possible positive influences and possible negative ones (ideas below).

Possible positive influences:

- Action- It could be argued that we wouldn't be aware of some of the biggest scientific issues needing tackling if it wasn't for the involvement of the media, and therefore the effect must be positive if we are becoming aware of pressing problems (e.g. obesity)
- Access - The media is a way through which the world of scientific research integrates with common people, who would not otherwise recognise the importance of science.
- Description – the media may be able to explain ideas in a more easily understandable manner, such as by identifying the key trends and emphasising the important facts, which makes science more understandable to those with less knowledge.
- It could be argued that any awareness of science is positive for those who would not have any interest otherwise.
- Examples of specific situations – e.g. in the banning of CFC sprays.

Possible negative influences:

- Simple facts may get blown out of proportion or taken out of context to such an extent that they could no longer even be considered factual or valid.
- False or flawed statistics are often given, taken from unreliable sources, which lead to confusion and the wrong message being portrayed.
- Oversimplification of ideas – everything is dumbed down and will not give the full story.
- Can create a biased public opinion – a conclusion may be reached which does not consider all particular explanations for a phenomenon.

Conclusion:

- Include a summary of all points made, and give a balanced overview of both sides of the argument.
- You may wish to reach a unanimous decision on the effect, making it clear that this is a personal opinion. Or you may wish to come to a compromise where the media are a positive in some situations or to some extent, but after a particular point or when taken too far, this effect can become negative.

3.   *"Why tell the truth if a lie is better for all concerned?"*
       In what circumstances can dishonesty be justified?

Introduction:

- It may be important to define the term 'lie' as a statement or implication that is directly opposed to the truth. It is also important to note that, in this context, a lie is a deliberate dishonest remark (not an accidental incorrect statement)
- You may wish to distinguish between lying as deliberately telling something that is factually incorrect, and deliberately avoiding or obscuring the truth.
- Introduce the idea that dishonesty is considered immoral and 'wrong' – the basic social rules are not to lie, children are taught from a young age that lying is bad behaviour.
- Begin to introduce you arguments in favour of lying in particular situations – the points you wish to explore later.

You may wish to structure this essay slightly differently to this mark scheme; through exploring a circumstance where it may be considered justifiable to lie, and then look at the alternative perspective arguing that it is not justified. Alternatively, it would be just as effective to make points for and against, as below.

Possible arguments for:

- In the case of 'white lies' – to avoid discouraging or offending people e.g. "Your hair looks nice"
- In self defence or when lying may save a number of lives – e.g. when people are taken hostage for their beliefs/ethnicity/race
- When talking to children about situations they wouldn't otherwise understand, especially when they ask difficult questions.
- To avoid an unnecessary, difficult conversation with someone who doesn't need to know something detailed – e.g. if they ask, "How are you?" not giving them a truthful answer may avoid a difficult conversation.

Possible arguments against:

- Creates a lack of trust and a lack of stability, feelings of guilt on both sides.
- Religious/foundations – 'do not lie' is one of the Ten Commandments.
- You could argue that lying will always cause more problems later down the line, even if making things easier in the short term.
- Advocating lying in some situations makes it easier to lie in other situations and creates a perpetual problem, it's impossible to draw a line other than to just say one should not lie as a rule.

Conclusion:

- Include a summary of all points made, and give a balanced overview of both sides of the argument.
- You need to reach a decision which illustrates if, and when, dishonesty can be justified. It's OK to also reach the conclusion that it never can be, so long as this is logically stated.

4. *"Science is a nothing more than a thought process"*
    What actually is science and how is it of value to us?

This essay is very open-ended: it is possible to take this question in any direction you feel appropriate depending on personal knowledge and interest so this mark scheme is by no means the only way of approaching this question – so long as a convincing and balanced argument is given and both elements of the question are considered, the answer will be credible.

Introduction:

- Include a definition of science – a systematically organized study or body of knowledge of the physical and natural world.
- Begin to introduce some of the key features of science.
- Begin to introduce some of the values that science has for society.

Possible points to consider – you want to contain a balance of defining and exploring the key features of science, and exploring the value of science within our society:

- Science is experimental – involving accurate and detailed study.
- Science could be argued to be physical and measurable- using known concepts in the real world that can be objectively identified.
- Science is adaptable and ever changing – according to research and current thought.
- Science could be considered to be held within the thoughts and brains of society – if there weren't people exploring it or interested in it, would it really exist?
- Science involves logical and critical thought – in that sense it could be considered to be merely a thought process – but you may wish to explore the idea that it couldn't exist without physical things to measure in themselves.
- Valuable for the evolution of society – as we learn more about the world and about ourselves, this helps change to occur which is vital.
- Provides concrete fact and stability.
- Gives a purpose to life, which brings fulfillment and contentedness.
- Brings explanation for some things that can't be explained through simple observations, which allows a firmer foundation and an answer to bigger, crucial questions.
- Helps us to gain more of an understanding of the world we are in – bringing a stronger grasp on reality and greater knowledge and insight.
- Medical reasons – drugs, medicines, health.
- Safety – e.g. monitoring volcanoes, predicting earthquakes etc.

Conclusion:

- Summarize the key ideas explored previously – the ideas associated with science and the values it may hold.
- Ensure to link back to the original statement – is it within our minds and merely a thought process – and does that diminish its' value?

END OF PAPER

Mock Paper B: Section 1

Question 1: C

This question has to be worked through in stages. To begin with, adjust his weekly pay to an annual salary, (560 x 52 = £29,120). Accounting for tax, his annual salary is 29,120/0.9 = 32,355. To account for the pay cut, 32,355 x 0.95 = £30,737. To deduct tax, subtract 20% of 20,737 (his taxable income) to give 4,147 tax. Subtracting the new tax from his new overall salary gives 26,589. Add on the new bonus of 90 per month to give £27,669, or £27,700 when rounded.

Question 2: B

The total saving on the final booking relative to the first is £230, but the cost of two cancellations must be deducted (£90) giving a total saving of £140.

Question 3: B

There are originally no odd numbered balls in Bag A. But as a result of the transfer, there could be an odd ball in Bag A. Therefore the probability of drawing an odd ball is found by multiplying the probability of selecting the new ball ($^1/_5$) by the probability that that ball is odd ($^2/_5$ – given by adding the one odd ball in the bag originally to the odd ball introduced) giving an probability of $^2/_{25}$ that the selected ball from Bag A is odd.

Question 4: B

Assume the price of bread is 100p. 100 x 1.4 x 0.8 = 112p after the subsidy. The cost of three loaves is therefore 336p (divided four ways this equals 84p per loaf)

Question 5: D

At 2120hrs, the minute hand is pointing to 4 and the hour hand is pointing one third of the way past 9 towards 10. $360°/12 = 30°$ – this is the number of degrees per hour division. Between the two hands then, there are 5 hour divisions plus an extra $^1/_3$. Therefore the angle is $(30x5)+(30/3) = 160°$

Question 6: B

There is a 3l and 5l bucket – therefore 4 litres can be measured from the difference between the buckets as follows. Fill the 5l bucket, decant 3l into the smaller bucket and then you are left with 2l in the large bucket. Pour this into the tank. Repeat the process again, decanting the remaining 2l into the tank once again to make 4l in total. The first time, 5 litres was required. The second time, the 3 litres from the second bucket could be tipped back into the 5l bucket, and then filled up with 2l of fresh water to then measure the final 2l in. Therefore 5 + 2 = 7 litres of water is sufficient to fill the tank with 4l.

Question 7: D

To answer this question, make a timeline showing the locations of the different genres of books. Place each book on the timeline as appropriate, making sure to indicate where more than one location is a possibility. From that, you will see that literature books are located to the right of engineering. This is true since they are to the right of art (which we know is right of mathematics (and therefore engineering, since the run between the sciences is uninterrupted)). The other statements, whilst potentially true, cannot be deduced for certain.

Question 8: C

The passage tells us that brand new cars lose value quickly, despite the car being virtually unchanged. Therefore in the absence of any contradictory information, it is reasonable to conclude that buying second hand cars is a wise choice.

Question 9: B

First, calculate how many bottles are sold. 2000 – (2000x0.9x0.8) = 560 bottles. Then divide the total profit by the number of units to give the profit per unit, which comes to 11200/560 = £20 per bottle.

Question 10: C

If ocean exploration has led to the discovery of many useful drugs, it could easily be said that it has benefitted many people in trouble. Whilst it might be cheaper than space exploration, the two are entirely different, and therefore people's views on the cost-effectiveness of space exploration cannot be directly compared to their views on ocean exploration. The other responses do not address the effect on people in trouble.

Question 11: E

The definition of timelessness requires something to be tested by time. Something that modern furniture cannot fulfil. Therefore statement E expresses a significant flaw in the reasoning. The other statements do not refer to the 'timelessness' aspect of furniture, therefore they are not directly relevant to the argument.

Question 12: A

The passage talks about the benefits of drinking red wine, not about living near to vineyards. The passage does not state that Italians drink more wine than Germans, therefore the assumption that they do is central to the argument.

Question 13: C

Tom arrives at 1620, and leaves 45 mins after Jane leaves. Therefore he also leaves 45 mins after Hannah leaves, since Jane and Hannah leave together. Since his journey is 10 mins faster than Hannah's, he arrives only 35 minutes after Hannah arrives (which happens to be 1620). Therefore Hannah arrives 35 minutes earlier than this, at 1545. Since she left at 1430, her journey took 75 minutes. Jane's journey took 40% longer (1.4 x 75 = 105 minutes). Therefore leaving at the same time as Hannah, 1430, Jane arrived 105 minutes later at 1615.

Question 14: B

This is a simultaneous equations question. Let x be the number of standard tickets sold, and y be the number of premium tickets sold.
Therefore: $x + y = 600$; $10x + 16y = 6,600$
$x = 600 - y$ » substitute: $10(600 - y) + 16y = 6600$
$6y = 600$; $y = 100$, therefore 100 premium tickets were sold.

Question 15: C

Between 20th January and 23rd April, there are 94 days. In 94 days, the moon makes $94/28 = 3.36$ orbits. This is equal to $3.36 \times 360° = 1210°$

Question 16: E

In question 14, you are looking for a strong opposition to the proposition that students at drama academies are not taught well academically. The strongest opposition would be evidence that such students perform academically well in some objective measure. Evidence of significantly above average GCSE results provides this.

Question 17: D

You should definitely draw this one out on paper. Trace out the paths and you find that both people have a net displacement of 11km to the North. Therefore since Anil is only net 2km East, and Suresh is 17km East of the starting point, there is a 15km separation between them

Question 18: E

If three times the final amount of concrete is ground off by the builder, three quarters of the original thickness is removed, hence one quarter remains. $14/4 = 3.5$cm

Question 19: A

Walking at 4mph, 3 miles takes ¾ hour = 45 mins. Adding the 5 minute stop, Chris will arrive at 1820, since he set off at 1730. At 24mph, 6 miles takes ¼ hour, 15 mins. Therefore setting off at 1810, Sarah will arrive at Laura's at 1825. Therefore Chris arrives 5 minutes earlier than Sarah.

Question 20: D

The passage tells us that illegal downloads are causing harm to the music industry. Whilst it gives an example, this does not mean the stated example is the principal issue. The conclusion that best fits the passage as a whole is to say illegal downloading is more harmful than many people think, given their willingness to undertake it.

Question 21: E

First, calculate the amount of water needed for each type of fire. Typical house fires require 20,000 litres, whereas garden fires usually need only 10,000 litres. Therefore all statements are correct EXCEPT E. Three house and ten garden fires require 160,000 litres to extinguish, not 140,000.

Question 22: E

Calculate the distance travelled during each component of the journey, then add them together. (20 x 30 = 600m, (30 + 20)/2 x 5 = 125m, 30 x 20 = 600m, 30/2 x 10 = 150m). Adding the distances together gives 1475m.

Question 23: D

The passage only talks about people's opinions on the scheme, and not about any action which could potentially be taken. Therefore the best summary is to say that more people oppose the scheme than support it.

Question 24: B + E

The question asks for two responses, therefore you must mark two and get them both correct for one mark. The suggestion is made that reducing fishing will improve fish populations. This assertion carries two major assumptions – that the fishing originally caused the decline, and that the decline is reversible, and can therefore recover if the threat is removed. Select these two responses for a mark.

Question 25: D

To calculate this, you need to work out how many possible combinations there are, and how many of them contain exactly two heads. Since there are 2 possibilities and 5 trials, the number of potential outcomes is $2^5 = 32$. For two heads, any combination of two coins can show heads – and since there are 5 coins tossed, there are 10 possible combinations of exactly two heads. Therefore the probability is $^{10}/_{32}$, which is equivalent to $^5/_{16}$.

Question 26: C

For a 5 litre bucket with a 2% margin for error, the maximum possible volume is 1.02 x 5 = 5.10l, and the minimum is 0.98 x 5 = 4.90l. Therefore there is a 200ml difference between the maximum and minimum volume possible. Therefore the range of cleaning powder required is 0.2 x 40 = 8g.

Question 27: B

To calculate the cost of the call, you need to first work out its duration in minutes and multiply by the off-peak rate per minute. Then you add on the connection fee. A call of 1.4 hours = 1 hour 24 minutes = 84 mins. (84 x 22 =1848p), adding the connection fee of 18p gives 1866p, or £18.66

Question 28: C

The passage tells us about the risks of sunbathing, and that many people do not see the danger in it. The final sentence shows us that the conclusion is a wide-ranging one, not a specific observation about UV radiation. Therefore Answer C is correct, it best sums up the passage as a whole.

Question 29: C

The sample argument gives a premise which represents a problem. There are two possible solutions, and since one is not available then the other solution is sought. Argument C has the same structure. The clothes need to be dried, and since the tumble dryer is out of action, the clothes are pinned on the washing line.

Question 30: C
First calculate the total pay, then divide this by the number of hours Jim works for an hourly rate. Total pay = 3 x 8 = £24; Total time = (11 x 8 x 2) + (15 x 7) = 281 minutes = 4.68 hours. 24/4.68= £5.12 [the total time is equal to the number of windows in total multiplied by the time taken to clean each window, plus the time travelling between the houses, which is 15 multiplied by the 7 journeys required]

Question 31: A
The passage argues that bottled water is pointless, as is almost identical to tap water. If bottled water had an additional benefit, such as being good for health, it might be that it makes sense to drink bottled water.

Question 32: A
Be careful – this sentence contains a triple negative. If the sentence read "...nor any cyclists that are marathon runners", it would be clear that no cyclists also run marathons. Changing the sentence to "...nor NO cyclists that AREN'T marathon runners" introduces a double negative, hence the meaning is not changed. Therefore it still means no cyclists run marathons, hence **A** is true.

Question 33: D
Draw this one out on a line. You will see that whilst we know Oakton is East of Langham, we cannot conclude its whereabouts in relation to Frampton – it could be either East or West. Therefore **D** cannot be said with certainty.

Question 34: C
Firstly calculate the surface area of the dome, then divide by the surface area covered by one pot to calculate the number of pots needed. A dome is half a sphere, so the area is given by $(4\pi r^2)/2$ = 12 x 49/2 = 294. Since one pot covers $12m^2$, 24.5 pots are required to cover the whole dome once and 49 to cover it twice.

Question 35: B
If 49 x 2 = 98 litres of paint, therefore 98 x 0.4 = 39.21 litres solid volume when the paint dries, = $0.0392m^3$. The volume of the hemisphere is $(4/3\pi r^3)/2$, dividing the reduction in volume by this gives 0.0392/686 = 0.0057%

Question 36: E
If the machine takes 400ms to wrap a sweet, it will wrap 10 sweets in 4 seconds = 0.4 seconds per sweet. Since there are 60 x 60 x 2 = 7200 seconds in 2 hours, and it can wrap 7200/0.4 = 18,000 sweets in 2 hours.

Question 37: C
If in 8 years time the sum of their ages is 52, the sum of their ages now is 52 minus two eights, which is 52 – 16 = 36. If they were both the same age, they would each be 36/2 = 18 years, however we are told John is 6 years older. Since the sum of their ages must still be 36, John has to be 21 and his brother is 15.

Question 38: B
The passage tells us that house prices have raised due to an increase in occupancy. Therefore making more houses available will make them more affordable. This could be achieved by either building more houses, or increasing the number of people in each house. Building more houses, C, is therefore a true conclusion based on information from right throughout the passage. The passage doesn't tell us what the current rate of house building is, it only speculates divorce as a cause of problems, and doesn't mention where divorcees live. Whilst it is true to say A, it is not the main conclusion of the passage, merely a paraphrasing of the first sentence, thus it is not the correct answer.

Question 39: C

If there are four stations between Crabtree and Eppingsworth, there are five independent journeys between them. Calculate the times between each of the stations, then sum them for the total journey time. The journey between stations 3 and 4 is leg 4 of the journey, and it takes 16 minutes. Therefore leg 5, the final component takes 16x0.8 = 12.8 minutes. Leg 3 takes 16/0.8 minutes = 20 minutes, leg 2 takes 20/0.8 = 25 minutes. Leg 1 takes 25/0.8 = 31.25 minutes. Summing these together gives:

31.25 + 25 + 20 + 16 + 12.8 = 105.05, about 105 minutes, since we do not measure rail journeys in fractions of minutes.

Question 40: E

The passage tells us residents fear the noise and disruption caused by the construction. Therefore E is the best answer, since it directly addresses the concerns the residents have. Residents would describe the other responses as unhelpful, since they do not address their main concern.

Question 41: D

To find out when the rental costs are equal, equate the two rental costs. If d is the number of days of hire, hiring from Tony's costs 23d, and hiring from Adam's costs 65 + 18d. Therefore at equality, 23d = 65 + 18d, 5d = 65, d = 13 days. However this is only at equality, and the question asks when a saving will be made by using Adam's. Therefore the answer is 14 days, when the first saving (of £5) is made by using Adam's.

Question 42: C

The passage tells us that antibiotic resistance could lead to people dying from Victorian diseases, and that liberal use of antibiotics in farming is the "most significant" contributor to this. Therefore it would be true to say that this use of antibiotics could cause serious harm.

Question 43: C

Only AMATAMA, option 2, is the same when viewed in a mirror and when viewed normally. The other options will not appear the same. To test this, you can use a shiny surface such as a mobile phone screen.

Question 44: B

You have to work through this sequentially. In week 2, 60% more purchases means 7,500 x 1.6 = 12,000 of purchases. A 30% profit margin means 1.3 x 12,000 = 15,600 of sales. In week 3, 2000 less in sales gives 13,600 in sales, and with a 60% profit margin, the purchase cost is 13,600/1.6 = £8,500.

Question 45: A

To answer this question, you need to work out the ratio of the areas. There are 6 squares and 8 triangles. If the area of a triangle is $3\sqrt{27}cm^2 = 0.5$ x base x height, $6\sqrt{27}$ = the value of the base x height. Here, you might see that $\sqrt{27} = \sqrt{(36 - 9)}$, therefore $(\sqrt{27})^2 = 36 - 9 = 6^2 + 3^2$, and applying Pythagoras' theorem, the equilateral triangle side length works out at 6cm. Therefore this is also the square side length. As a result, the total area of the squares is $6 \times 6^2 = 216$. The total area of the triangles is $8 \times 3\sqrt{27}$. Therefore the total surface area is. $216 + 24\sqrt{27}$. The area of one square, which we are interested in, is $6^2 = 36$, so the probability of it landing on that portion of the area is $36/(216 + 24\sqrt{27})$. However this is not the simplest form. Factorise the denominator to $36/9(24 + \sqrt{27}) = 4/(24 + \sqrt{27})$.

Question 46: B

The passage tells us that the median length of commute is 40 minutes: therefore at least 50% have a commute which is equal to or shorter than 40 minutes, therefore B is true. Option E is not true, as the passage tells us that over 50% of *commuters* complain about the duration of their commute – but since not all people commute to work this might not be more than half of the population.

Question 47: C

To solve this, calculate the number of possible outcomes ($3^3 = 27$) and the number of outcomes which contain no score. In every theoretical round, one third of outcomes have no score. Therefore after round one, there is a $1/3 = 9/27$ chance of no score. In addition, in the next round there is a $1/3$ chance of the other $2/3$ being zero, and in the third round there is a $1/3$ chance of the remaining $2/3$ being zero. Therefore the probability of at least one zero score is $(1/3) + (1/3 \times 2/3) + (1/3 \times 2/3 \times 2/3) = 9/27 + 6/27 + 4/27 = \underline{19/27}$.

Question 48: C

The graph shows that more water is required around the middle section for each additional increase in depth, with less at either end. Therefore the container is spherical.

Question 49: C

The passage states only problems with owning a listed building, and not a benefit. Therefore C is true – it is not a balanced argument, and there could be benefits to owning such a house that the passage makes no account of, therefore it might not be, on balance, bad to own such a property.

Question 50: B

The answer is found by an iterative process, multiplying each intermediate answer by the percentage decrease in a compound interest deduction. $36000 \times 0.75 = 27,000$. Keep taking ¾ off, repeating ten time to give the answer of 2,030 cars a decade later in 1015.

END OF SECTION

Mock Paper B: Section 2

1. *Design an experiment to deduce the sensitivity of a snake's hearing. Explain everything you would do, and your rationale for doing so.*

This question is different to most others you will encounter and does not demand the usual structure. Despite this, there should be a logical flow to the answer, explained comprehensibly and clearly, and reasoning should be justified adequately to provide support.

Introduction:

A brief overview of the independent and dependent variables would be appropriate, as well as a short hypothesis about what you may expect to observe.

Points to consider and justify:

- Independent variable = what you will change, probably the frequency or volume of the sound. You will be required to operationalize this variable, considering the equipment used to provide the sound – such as a computer, and the method of delivery to the snake – through speakers, through headphones etc.
- Dependent variable = what you will measure, to do with the snake's response to the sound. This could be operationalized in a number of ways, such as a detectable movement, which has been previously learnt by association, or simply a twitch – but this should be scientific and objective. Any equipment or materials, including an observer, should be identified.
- Sample and repeats = how many different snakes will you test? Will you repeat the experiment on the same snake a number of times?
- Setting = is this a laboratory experiment or will it be taking place in a naturalistic environment?
- Control variables = it must be clear that you have considered potential confounding factors and have taken steps to control them. These may be extraneous noise (control for by taking place in a soundproof room), errors in determining when the snake has heard the noise (repeats are probably the easiest way of eliminating this), sample bias – are you testing just one breed of snake or lots of different ones? Observer bias – if an observer is detecting movement, ensure they are doing this reliably.
- Interpretation of results – include how you will present your data and suggest any data manipulation techniques (graphs, statistical tests) outlining what you hope this will achieve.
- Reporting the experiment – explain how you might wish to write up the experiment and any further research you may want to highlight.

Ideas for justification:

- Convenience
- Time taken
- Accuracy
- Validity
- Reliability
- Access/understanding

Conclusion:

Summarise the experiment in a couple of sentences AND your overall justification about what you wish to achieve, identifying both the positives and potential limitations or need for further study.

2. "The eternal mystery of this world is its comprehensibility"

To what extent is the world comprehensible?

Introduction:

- Include a definition of comprehensible – understandable, intelligible
- You may wish to consider the extent to which comprehensibility in itself is subjective, and therefore your essay will be biased towards your own views on comprehensibility.
- Introduce a summary of the ideas you want to explore within the essay on both sides – arguments for the world being comprehensible and arguments against (see ideas below)

Points for comprehensibility:

- We are restrained by the constraints of our own human mind – is the question how much can the human mind comprehend? The world is limitless until we reach the limits of our own minds.
- We are still discovering more and more about the world and haven't reached our limit yet – there is always more to discover and further our knowledge and understanding. Until we can no longer advance in our scientific understanding, the world is comprehensible and in our hands to discover.
- One could argue that we are the most advanced and evolved organisms in the world and therefore everything we need to understand must be less complicated than ourselves.
- With an understanding of history, we can understand why things are the way they are – as long as we learn from the past, surely there is no more in the future we cannot reach?

Points against comprehensibility:

- One could argue that the world is contained within our minds, and since we are within our minds we cannot objectively understand them – do we have to be outside of the world to comprehend it, and therefore not be of this world?
- The human mind may be so complex that it is, paradoxically not complex enough to understand itself.
- There must be a limit to our understanding, since the world has limits and our brain is a limited place for processing.
- We do not know what else there is to comprehend, and therefore we will never know when we have reached the limits of our comprehensibility – therefore never fully reaching complete understanding.
- Would we be God if we could comprehend the world and everything in it?

Conclusion:

- Include a summary of all points made, and give a balanced overview of both sides of the argument.
- Reference back to the original quote – it may be relevant to emphasise the word 'mystery' in the context.
- Draw together your points in some kind of concluding statement – either to say the world is completely comprehensible or to say that it is only comprehensible to some extent would be justified, so long as this is logically and critically summarised and justified.

3. "The greatest obstacle to learning is education"

Argue for or against this statement.

Introduction:

- Consider a definition for learning – the acquisition of knowledge; versus education – a structured approach to transferring information about the world from one person to the other.
- Identify the ways through which knowledge can be acquired – through teaching, experience, study – emphasise that learning doesn't have to be purposeful, whereas education involves the intention of learning (whether or not it is achieved).
- This essay asks for ONE perspective – introduce the argument you wish to pursue.

Possible arguments for:

- The greatest feats of learning could be considered the acquisition of language and/or movement – all of which are done without and before education.
- Education is directive – based on the views of a few people about what they believe is important –and therefore it narrows our minds to only the concepts taught.
- Everyone learns in different ways – creatively, actively, through vision, through hearing and education, as we know it is too structured to suit everyone.
- School is a bubble – set apart from the real world – you could argue that the real learning happens once children leave school and have to fend for themselves.
- Have homeless people or those with no access to education learnt less? They may have learnt different skills, but broader, more practical ones.
- Learning involves actively engaging with the information given, and requires an understanding – you could say that education teaches simply recall of facts, not true knowledge of the world.
- Examples of people who have achieved much without a formal education e.g. Alan Sugar

Possible arguments against:

- You could say that education is feeding the natural curiosity within human beings – we naturally want to learn more about the world and formally educating people of this is an easy and natural way to pass on knowledge.
- Education provides not just academic knowledge, but social knowledge; how to behave, what is right and what is wrong, how to make friends. There isn't such a concentrated opportunity to learn these things in other places.
- Reading and writing are basic skills required for so many careers – education opens the doors to further learning
- How could one say that a person has learnt nothing after graduating from school? Therefore is cannot be a barrier to learning, because children surely learn something!
- Knowledge needs to be passed on to those capable of furthering it – and therefore education provides a means of passing on knowledge and developing our understanding of concepts – if education didn't exist, development would be slower so we must be learning something.
- Uneducated people make up the majority of our unemployment figures and are often more likely to turn to drugs, alcohol and develop other social problems.

Conclusion

This is an argumentative essay so the conclusion MUST reach a decision. Summarise the key ideas of the essay and dismiss any opposing perspectives.

4. Does a vacuum really exist?

Introduction:

- This must include a definition of vacuum – a space entirely devoid of anything.
- You may wish to clearly indicate what you mean by 'anything' – matter, time, etc.
- This could be considered philosophically or physically – make it clear which (or both) perspective you are wishing to take.
- Introduce the key ideas on both sides you wish to pursue on both sides of the argument – reasons for it existing and reasons against (see below)

Arguments for:

- How can we prove that 'nothing' exists – in proving there is nothing, there must be something, in order for there to be something to prove.
- Just because our brains cannot comprehend a place where there is nothing, that does not mean that it doesn't exist – we just may not be adequately equipped to understand it.
- We can never prove that a vacuum doesn't exist, because there is no way of finding it if there is nothing there – so we must assume it exists if we cannot disprove it.
- If there is a place where there is something, there must also be a place where there is nothing, in order for the place where there is something to be valid.

Arguments against:

- Without matter, there is an absence of anything, and therefore without anything, there is nothing – so it cannot exist.
- There are fields and properties and relativity everywhere in the known universe, provided the physics applies equally everywhere, which means there is always something there, so there is nowhere in the universe that there is nothing.
- In order for something to exist, it must have properties which can be defined and demonstrated within the realms of our reasoning – since this is not the case for vacuums, they may not exist.
- Our understanding of physics is by no means complete and therefore our ideas about what matter is and how to define it are likely to change – therefore there is no reason to believe there must be a place where there is nothing – it could just be filled by something we haven't discovered yet.

Conclusion:

- Draw together all the points made on both sides of the argument – summarise the key points and bring them down to earth again.
- The conclusion must be solid and clear – even if not particularly complicated, it must be logical and understandable, in order to bring together this potentially confusing subject.
- You may wish to reach a decision, or decide that it depends on the perspective you take – as long as this is correctly justified.

END OF PAPER

Mock Paper C: Section 1

Question 1: D

Answer C) is completely irrelevant, so is not a flaw. Answer B) is not a flaw because when assessing an argument, anything that is stated (i.e. not concluded from other reasons in the passage) is accepted as true. We do not require evidence or sources for any statistics presented. Answers A) and E) are both claiming that something is immoral, which is thus expressing an opinion on the part of the arguer. This is not a flaw, the arguer is at liberty to claim something is immoral, and to claim that the government is morally obliged to act, and that it has not done so. However, answer D) identifies a valid flaw. The argument rests on us accepting that if there were less uninsured drivers, there would be less crashes. This is not necessarily correct, so D) is a flaw in the passage.

Question 2: A

The passage does not say anything about whether Brazil should legalise guns, it simply reports what one commentator *said* was the reason why the move to ban guns was unsuccessful. Thus, C) and D) are not best supported or opposed by the passage as directly as A) and B). The passage clearly indicates that the UK *should not* legalise guns, when it says "legalising ownership in the UK would be a bad move". Thus, A) is best supported by the passage. Since A) is best supported, E) is also incorrect, and the answer is A).

Question 3: C

For each of the walls where there is no door, the wall is 6 tiles high and 5 tiles wide, which is 30 tiles. The wall where the door is requires a row of 2 tiles above the door, then there is a width of wall of 120cm which requires completely tiling, which is 6 tiles high and 3 tiles wide, hence this wall requires a total of 20 tiles. Hence a total of 110 tiles are required for the walls. The floor is 2 metres by 2 metres, so 5 tiles by 5 tiles, hence 25 tiles are required for the floor. Hence the answer is 135.

Question 4: D

Answer E) is irrelevant to which of Trevor and Jane will arrive first, so does not weaken the conclusion. Answers A), B) and C) all strengthen the answer, giving further reasons why we might expect Trevor to arrive first. Answer D), however, would slow Trevor down, meaning that it was more likely that Jane would arrive first. Thus, Answer D) weakens the passage's conclusion, and hence Answer D) is the answer.

Question 5: A

He has enough butter to make 2.5 times as many cupcakes as the recipe, which is 50
He has enough sugar to make 3 times as many cupcakes as the recipe, which is 60
He has enough flour to make 5 times as many cupcakes as the recipe, which is 100
He has enough eggs to make 3 times as many cupcakes as the recipe, which is 60
The lowest of these is 50, so he can make 50 cupcakes. He needs 2.5 x 4 eggs to do this, which is 10 eggs. Therefore he has 2 eggs left over.

Question 6: B

Let the number of minutes that the journey takes be 't'. So ABC charge 400+15t pence for the journey. We can calculate that XYZ taxis charge 400+(30x6) = 580 pence. In order for both journeys to cost the same, 580=400+15t. 180=15t, thus t=12. Therefore the 6 mile journey needs to take 12 minutes. 6 miles in 12 minutes is 30 miles per hour, so the answer is B.

Question 7: E

We can see that all of answers A) through D) are essential for the conclusion to be valid from the squire's reasoning. Lancelot must have great courage, this must be a requirement for the Adzol, and no other knights must have sufficient courage, in order for us to be certain that Lancelot will succeed but all of Arthur's other knights will fail. Thus A) and B) can be clearly identified as assumptions. C) and D) require a bit more thought, but we can see that nothing in the passage explicitly states the Elders' tales are correct. If the elders are not correct, then great courage may not be required to be successful in the Adzol. Thus, both C) and D) are also assumptions. Hence, the answer is E).

Question 8: E

B) is incorrect, as the passage does not say that arch-shaped gaps *always* indicate where windows once stood, simply that *these arches* do. C) is also incorrect, as the passage simply states that windows are not found in *underground halls*. A) is a reason in the passage, and is not a conclusion. D) and E) could both be described as conclusions from this passage, but we see that if we accept D) as true (along with the fact that the hall is now underground), we have good reason to believe that E) is true, whereas E) being true does not necessarily mean that D) is true. Thus, E) is the *main* conclusion, whilst D) is an *intermediate conclusion*, which supports the main conclusion.

Question 9: D

Usually bread rolls cost 30p for a pack, but if the cost per bread roll is reduced by 1p then a pack will cost 24p. Hence we need to find z, where $24(z+1)=30z$, where z is the original number of packs that could have been afforded. $24z+24=30z$, hence $24=6z$, so $z=4$. Hence he was originally supposed to be buying 24 bread rolls.

Question 10: B

Answer E) is an irrelevant statement that says nothing about whether England *do* have good players. Answers A) and D) actually weaken the sporting director's arguments, suggesting that England may have a good team, and it may just be poor performances in world cups, and not a lack of talented players. This leaves B) and C). C) may appear to strengthen the sporting director's argument, but on closer inspection we see that in fact it says that for the last 70 years, England have had at least 1 player in the top 10 in the world. This does *not* strengthen the argument that England have been lacking talent for the last 25 years, and may actually reinforce the chairman's argument that it is simply the *current* crop of players that are not good enough. Answer B), however, does strengthen the argument, suggesting that England's performances have been poor over the last 20 years, thus strengthening the argument that there may be a lack of talented players that has been ongoing for a couple of decades, as claimed by the sporting director.

Question 11: A

He can prepare each batch of cakes while the previous one is in the oven but it takes longer so we have to allow 25 minutes for each batch, plus 20 minutes for the last batch to cook while no further batch is being prepared. There are 12 in each batch, so for 100 cupcakes there needs to be 9 batches. Hence the total time needed is 25 minutes x 9, + 20 minutes. This is 245 minutes, or 4 hours 5 minutes. Hence to be ready by 4pm he needs to start at 11:55am, so the answer is A.

Question 12: D

We can first work out the rate of girls' absenteeism. First we need to work out how many of the pupils at Heather Park Academy and Holland Wood Comprehensive are girls. Let g be the number of girls in Heather Park Academy. Then $0.06(g)+0.05(1000-g)=(1000)(0.056)$. Then $0.06g-0.05g=56-50$. Then $0.01g=6$, so $g = 600$. Hence 600 pupils at Heather Park Academy are girls. The proportions at Holland Wood Comprehensive are the same but there are half as many pupils, so 900 pupils at the two schools combined are girls.

The average absenteeism of girls is 6.1%. We know that 900 of the 1100 girls have an average absenteeism rate of 6%. Let the average absenteeism rate of girls at Hurlington Academy be r. Then $900 \times 0.06 +200r = 0.07 \times 1100$. Hence $54+200r=77$. $77-54 = 200r$. $23/200 = r$. $r=0.115$. Hence, the rate of absenteeism amongst girls at Hurlington Academy is 11.5%

Question 13: C

A), B) D) and E) are all directly stated in the passage, so can all be reliably concluded. Perhaps the trickiest of these to see is answer D), which is true because the passage says "*due to*" the advent of more accurate technology, thus clearly identifying that this had *caused* the switch to the situation of most watches being made by machine. C), however, is *not* necessarily true. The passage states that most *watches* are produced by machines, but only states that *some* watchmakers now only perform repairs. This does not necessarily mean that most watchmakers do not produce watches. It could be that only a handful are required in the entirety of the watch industry for repairs, and that the numbers still producing watches exceeds those in the repair business. Thus, C) cannot be reliably concluded from the passage.

Question 14: A

B) is not a valid conclusion from the passage, because the fact that someone uses an illogical argument (as some Pescatarians are claimed to in this passage) does not mean that they cannot use logic. D) and E) are not conclusions from this passage because the passage is not saying anything about the ethicality of eating meat, but simply commenting that one argument used against doing so is not logical. Answers C) and A) are both valid conclusions from the passage, but we see that if we accept C) as being true, it gives us good cause to believe that A) is true, but this does not apply the other way round. Thus, C) is an intermediary conclusion, whilst A) is the main conclusion.

Question 15: E

The research conducted does not ask about whether it is *important* to learn some of the language before travelling abroad, simply whether participants *would*, so B) cannot be concluded. D) is incorrect because the passage states *15%* would, which is clearly not less than 10%. The passage states that this is symptomatic of a deeper underlying issue, but does not say that many issues of racism stem from this, so C) cannot be concluded. Now, the passage states that 60% of people feel foreign people should learn English before travelling to Britain, and 15% of people would attempt to learn the language before travelling to a country which did not speak English. However, this 15% could be some of the same people as the 60%, in which case A) would be incorrect. Thus, A) cannot be reliably concluded. However, there must be at least 45% of people who feel that foreign people should learn English, but would not learn a foreign language themselves, so E) *can* be reliably concluded.

Question 16: D

She needs to print 400 x 2 = 800 double sided A4 sheets, which will cost £0.03 each. Hence the total cost of this is £240. She also needs to print 1500 single sided A5 sheets, which will cost in total £150. Hence the total cost is £390.

Question 17: B

The passage has stated that if Kirkleatham win the game they will win the league, so E) is not an assumption. Meanwhile, the manager has stated A), C) and D), and the passage has not claimed anything about whether Kirkleatham can easily win the game, so A) and D) are not assumptions. However, B) does identify an assumption in the passage. The fact that Kirkleatham will not win the game without playing with desire and commitment does *not* mean that they will win the game if they do play with desire and commitment. And we can see that for the argument's conclusion (that Kirkleatham *will* definitely win the league) to be valid from its reasoning, this is required to be true. Thus, B) identifies an assumption in the passage.

Question 18: B

Answers A) and E) are not relevant, because neither affect the strength of the councillor's argument from a critical thinking point of view. The councillor's argument says nothing about house prices, simply the cost of building the estate and the effects on Wildlife, so A) is not relevant. E) is not relevant because additional support, or likelihood that it will be heeded, does nothing to affect the strength of a given argument. C) and D) actually strengthen the councillor's argument, suggesting that brownfield land does have good infrastructure (C)) and that the greenbelt areas do have a lot of wildlife (D)). B) does weaken the councillor's argument, as it suggests that building on brownfield land may also have adverse impacts on wildlife.

Question 19: B

The reasoning in the passage can be summarised as "IF A happens, B WILL happen. IF B happens, C WILL happen. Thus, if A happens, C will happen". Only B) also follows this reasoning. A) can be summarised as "If A does NOT happen, B will happen. A doesn't happen, so B will happen". C) can be summarised as "If A happens, B will happen. Thus, if A happens, C can happen". Here, there is no explanation of "If B happens, C will happen", so it is not the same reasoning as in the question. D) reasons as "If A happens, B will happen. If A doesn't happen, C will have to happen. Thus, to prevent B happening, C will have to happen", which is vastly different to that in the question. E) is the direct opposite of the question, reasoning as "If A doesn't happen, B can't happen. If B doesn't happen, C can't happen. Thus, if A doesn't happen, C can't happen".

Question 20: B

We can tell the amounts for the green party and the blue party are both 1/3 of the total, and that the amount for the red party is 1/4 of the total. Hence 1/12 is left, so the amount for the yellow party must be 1/12. Hence the red party have 3 times the intended vote of the yellow party.

Question 21: B

A large pizza with mushrooms and ham is £12, garlic bread is £3, chips are £1.50 x 2 = £3, a dip is £1, hence the current total is £19. The cheapest way to order this is to get the price up to exactly £30 as this will reduce the price to £18. This takes £11. Only one of these options costs £11, which is a large pizza with mushroom. Hence the answer is B.

Question 22: D

The reasoning in the passage follows the style: "If A doesn't happen, B can't happen. If B doesn't happen, C will happen. Thus, if A doesn't happen, C will happen". Only answer D) follows this style. Answer A) simply reasons as "If A happens, B will happen. A doesn't happen, so B won't happen", which is both incorrect and a different style from the question. Answer C) is also incorrect, reasoning as "A is being considered in order to do B. If B happens, C will happen. However, A does not happen, so C won't happen", again both incorrect and a different style to the question. Answer E) follows the pattern "If A happens, B won't happen. If B doesn't happen, C will happen. A happens, so C will happen". Answer B) is similar to answer D), but not quite the same. Answer B) follows the pattern of "If A does not happen, B will not happen. If B can't happen, C can't happen. Thus, if A doesn't happen, C can't happen". In answer B) and the question itself, C *will* happen if A doesn't happen, so there is a slight difference in Answer B). Hence, the answer is D).

Question 23: D

The question simply describes how a combination of factors are responsible for the M1 Abrams being the world's most formidable tank, so the view of country X is incorrect. It does *not* claim that it is impossible for a tank to be as good as the M1 Abrams, so E) is not a valid conclusion. Equally, it does not say the new tank's armour will not be as good as the Abrams (in fact it is implied that it may well be as good), so C) is also incorrect. A) B) and D) are all valid conclusions from this passage, but we can see that A) and B) contribute towards supporting the conclusion in D). Thus, D) is the main conclusion of this passage, whereas A) and B) are *intermediate* conclusions given to support this main conclusion.

Question 24: D

We can simply add up the amounts in the bank accounts and find the difference between each month – it doesn't matter that the salary is paid in as it is the same every month. Doing this, we find out the biggest difference is between 1st May and 1st June, hence the answer is D, May.

Question 25: B

Firstly we can work out the full dose for the son. He needs to take 0.2ml per kg of weight for each dose, and he is 40kg, so this is 8ml. He takes 40 doses altogether, so in total he needs 320ml of medicine.

Then we work out the doses for everyone else, add them together and halve them. The daughter's full dose would be 2ml, 30 times, which is 60ml altogether. My dose would be 7.5ml, 60 times, which is 450ml. My husband's dose would be 8ml, 60 times, which is 480ml. Altogether, this is 990ml. However, only half of these dosages is needed, which is 495ml. Hence the total needed is 320 + 495, which is 815ml. Hence 5 200ml bottles of medicine are needed for the full course.

Question 26: A

There are 21 forks and 21 knives. If half as many are red as blue, and half as many are blue as yellow, they are in the ratio red:blue:yellow 1:2:4. Hence of the 21, 3 are red, 6 are blue and 12 are yellow. Hence the probability of getting a yellow knife is 12/21 = 4/7. The probability of getting a red fork is 3/21 = 1/7. Hence the probability of getting both is (1/7) x (4/7) = 4/49.

Question 27: D

The Principle used in the passage is that public funds raised through taxation (which is compulsory) should not be used for any services unless they benefit everyone, such that nobody is forced to pay for services that do not benefit them. Answer D) is the best application of this principle, as it directly follows it. Answer B) mentions public funds being used to support a service that benefits the whole country, but this does not necessarily mean that they *shouldn't* be used to support services that don't benefit everyone, so answer B) is not as directly an application of the principle as answer D). Answer E) is not the same principle because this is talking about funds being used for services that benefit the *country*, rather than everyone in it. Meanwhile, Answer A) is talking about how many people *use* a certain service, rather than how many people *benefit* from it, so this is not the same principle. Answer C) is completely different, talking about funds being used because some cannot afford private health service, regardless of how many people are benefitting from the public health service.

Question 28: B

The chairman has *stated* that 'All Inclusive' services are more popular than 'Hourly' services. He has not deduced this from any evidence, and thus he has assumed nothing about their popularity. Thus, C) and D) are incorrect. The chairman's argument is simply that focusing on 'All Inclusive' services will bring in more profit than 'Hourly' services, as he says they should focus on 'All Inclusive' *rather* than 'Hourly' services. Thus, any reference to other services or other profit-raising strategies are irrelevant, so A) and E) are irrelevant. However, B) correctly identifies the chairman's flaw. Just because 'All Inclusive' is more *popular* than 'Hourly' services does not mean they are more *profitable*, and if they are not then the chairman's conclusion is no longer valid. Thus, B) correctly identifies the flaw in his argument.

Question 29: A

B) is not an assumption because the passage *states* that renewable sources do not cause damage, so we accept this as true. E) is not a flaw, because again the passage has stated that the use of these fuels to produce power will continue to cause climate change *as long as it continues*, thus we must accept as true that it cannot be halted or prevented whilst these fuels are used. C) and D) are irrelevant to the argument's conclusion that if we wish to stop damage to the environment, we need to switch to renewable fuels, and thus they are not flaws. However, at no point is it stated that *all* non-renewable fuel sources cause environmental damage, it is only stated the non-renewables *such as* Oil, Coal and Natural Gas do.

Thus, we have no guarantee that Nuclear fuel will cause environmental damage, and if it doesn't, the passage's conclusion no longer stands. Thus, A) is a valid assumption in the passage.

Question 30: E

Answers A) and D) do *not* strengthen or weaken the argument because the question states that this increase in non-vaccinated individuals has occurred despite powerful evidence of vaccine safety, and in spite of advice from doctors. This suggests that people who do not vaccinate pay little attention to evidence of advice from doctors, so we should not expect these factors to have much of an effect. B) is completely irrelevant to whether the rate of increase will continue. C) actually weakens the argument, suggesting that such increases are common, and normally stop after 6 years. If the current increase was to follow suit, it would stop next year and vaccination rates would not fall below 90%. E), however, implies that this kind of increase has happened only once before, and in this case it continued for 13 years. If the current increase was to follow *this* pattern, it would continue for another 8 years, where vaccination rates would be below 90%. Thus, E) strengthens the argument's conclusion that we should expect an outbreak of measles.

Question 31: A

Answers B) and E) are both in contradiction with stated points in the question, which states that there is now powerful evidence human bodies *are* set up for long-distance running, and that it is well established that humans evolved in Africa. Answer C) is irrelevant, because the presence of other theories does *not* necessarily affect whether we should believe this theory based on the new evidence. Answer D) actually strengthens the argument, suggesting that the new evidence does provide powerful reasons to believe this theory. Answer A) however is a valid flaw, that evidence supporting a theory does not necessarily *prove* that it is true. Thus, the answer is A).

Question 32: C

The percentage of students who had their grades predicted correctly is the same as the number who had their grades predicted correctly as there are 100. Hence we simply need to add up the numbers on the diagonal of the table, where actual grade is the same as predicted. This adds up to 39, hence the answer is C.

Question 33: B

The 2 wage reductions mean that when the wage increases happen, the raises will be x% of a smaller number than the decreases were. Thus, the wage will not rise as high as the original level.

If you are struggling to visualise this, the easiest way to do it is to substitute a number for x. Let us do the calculation treating x as 10.

The first wage drop is by 10%. Thus, the wage is now 90% of the original wage.

The second wage drop is also by 10%, but at this point, the wage is only 90% of the original wage. Thus, the drop will be by 10% *of 90%* of the original wage, resulting in a new wage of 81% of the original wage (10% of 90% is 9%)

Then, we have the first increase, which will be 10% of this new wage (81% of the original wage). Thus, after the first increase, the new wage will be 81% +(10% of 81%) of the original wage. Thus, it will be 81% + 8.1%, which is 89.1% of the original wage.

Now we have the second increase. Another 10% is added, this time of 89.1% of the original wage. We now have an increase of 8.91% (10% of 89.1). Thus, after the final raise, the wage will be 89.1% + 8.91%, which is 98.01% of the original wage. Thus, the new wage is lower than the original wage.

Question 34: C

The passage says nothing about whether it is more important to have good doctor-patient relations than scientific progress, so answer E) is not a valid conclusion. Answer D) is a direct argument against the question, which claims the naming system *should* be changed, so answer D) is not a conclusion. The passage states that the confusing system causes problems in scientific literature, but this does *not* necessarily mean that changing it would allow faster progress in scientific research, so answer B) is not a valid conclusion. Answers A) and C) are both valid points from the passage, but we can see that Answer A) is a *reason* stated in the passage, which helps support the statement in C). Thus, Answer C) is the main conclusion.

Question 35: E

180 people applied in total. 128 did French, 64 did Spanish, 48 did neither French nor Spanish. We hence know that exactly 132 people did either French, Spanish or both. Since 128 people did French, only 4 people can have done Spanish without doing French. Hence 60 people must have done both French and Spanish.

Question 36: E

Answer A) directly strengthens the argument, suggesting that increased funding on education cannot cause a larger reduction in drug usage. Answers B) and D) do not directly affect the argument either way. Both these answers concern whether providing youth centres/recreational activities in areas of high social deprivation can cause reductions in drug usage. However, this does not directly affect the argument's conclusion that education alone cannot provide a further reduction in drug usage. Answer C) also does not affect the answer, for similar reasons, as it does not necessarily affect whether education can or cannot provide a further reduction in drug usage. Answer E), however, directly weakens the argument, suggesting that Spain has achieved a greater reduction in drug usage than the UK, simply by providing education. This directly challenges the argument's conclusion that education alone cannot further reduce drug usage.

Question 37: D

During June and July there are 61 days. The show will have to travel 8 times in the two months, taking 2 days each time, so there can be a maximum of 61 – 16 = 45 days of performances. Given the musical spends the same number of days in each place, this means 5 days of performances in each location. Given there are 5 days of performances and 2 days of travelling, which totals a whole week, the show must always travel on the same days of the week and perform on the same 5 days. The best time of the week to travel is Thursday and Friday because this means missing no matinee performances, so the most performances happen when each location's performances start on a Saturday and finish on a Wednesday. This means there are 8 performances in each location, a total of 72 performances. Hence the maximum number of people that can see the show is 72,000.

Question 38: A

If every swimmer swims 3 identical times, then each swimmer will have 3 consecutive ranks. Hence the best swimmer will have swims 1, 2 and 3, the second will have 4, 5 and 6, etc. Hence the first swimmer's total score is 6, the second's is 15, the third's 24 (7+8+9), the fourth's 33 (10+11+12), the fifth's 42 (13+14+15), the sixth's 51 (16+17+18), the seventh's 60 (19+20+21), the eighth's 69 (22+23+24), the ninth's 78 (25+26+27), the tenth's 87 (28+29+30). Hence 9 swimmers proceed to the final as the tenth best swimmer has a score of over 80.

Question 39: B

The question states A), B), C) and D), so all of these can be reliably concluded. However, the passage does *not* state B). The passage claims that *most* of the public in Europe is scared of nuclear power, and that some point to the disasters mentioned as evidence that it is not safe. However, this does not necessarily mean that these disasters have caused these fears. It could be a minority using these as evidence, and that the rest of the public has other reasons for fearing nuclear power.

Question 40: C

If Megan's mixture contains 3 times as much water as squash, then it must contain 9 litres of water and 3 litres of squash. If bottles of squash contain 1 litre to the nearest decilitre then they can contain as little as 950ml. Hence to guarantee she has 3 litres of squash, Megan must buy 4 bottles of squash.

Question 41: A

Andrew runs 3/4 of the distance that Alice does in the same time, therefore he must run at 3/4, or 75%, of the speed that Alice does. Amanda runs at 60% of the speed that Alice does. Hence Amanda runs at 60/75, 4/5 of the speed that Andrew does. Putting it a different way, Andrew runs 1.25 times quicker than Amanda. If Amanda runs at 200m per minute, then Andrew will run at 200 x 1.25 = 250m per minute. 10000/250 = 40, hence Andrew takes 40 minutes to run the race.

Question 42: E

Although it comes at the beginning of the passage, E) is actually the main conclusion. We can see that if we accept B), C) and D) as being true, along with the fact that the NHS needs more funds to deal with the extra workload caused by the ageing population, then we have good reason to believe E). Thus, E) is the conclusion, and B), C) and D) are all reasons supporting it. A), meanwhile, is a counter argument, which is then refuted in the passage.

Question 43: A

England won Pool C so they will be in Quarterfinal 3, where they will play Brazil. If they win, they will play the winner of Quarterfinal 1. Hence they can only meet teams from Quarterfinals 2 or 4 in the final. These teams are Argentina, Nigeria, South Africa or Holland. Hence the only one of these 5 teams they can play in the final is Nigeria.

Question 44: D

To make 1 metre she needs 4 of the 300mm length or 5 of the 210mm length. There are hence 3 joins of length 210x5 = 1050mm, totalling 3150mm, and 4 joins of length 300x4 = 1200mm, totalling 4800mm, on each side. Hence on each side there is 5850mm of tape. Hence the total tape is 10,700mm or 10.7 metres.

Question 45: E

The team must have played 12 matches at home and 12 away. If they have won 18 matches altogether and have won twice as many matches at home as away then they must have won 12 matches at home and won 6 matches away. They have hence lost no matches at home and 6 matches away. Therefore the answer is 6.

Question 46: C

The passage has *stated* that the fear is illogical, so E) is not a valid assumption. Meanwhile, whether there are other contributing factors is irrelevant to whether sci-fi has caused some cases of public fear of scientific progress, and the passage has not assumed it is the sole reason, so D) is not a valid assumption. B) actually supports the passage by reinforcing one of its reasons (how the viruses in many sci-fi movies are impossible), whilst A) is not a flaw because the passage states how *many* sci-fi movies are responsible for causing illogical fear of science. However, the passage has assumed that deaths could be prevented without the hindrance to progress from sci-fi movies, and this does not necessarily follow on from its reasoning. Thus, C) is a valid assumption from this passage, and C) is therefore the answer.

Question 47: B

Ashley walks 8km at 8km an hour, therefore he takes 1 hour to get there, so he leaves at 12pm.
Ben gets the bus, which takes 40 minutes and departs at 25 minutes past or 55 minutes past. If he gets the 12:25 bus he will be late so he needs to leave at 11:55am.
Callum cycles 12.5km at 12km an hour, which will take him 62.5 minutes, so he needs to leave at 11:57:30am.
Dave gets the train which takes 20 minutes and goes every 10 minutes, so the earliest he will need to leave is 12:20pm.
Ed gets the park and ride bus which comes every 10 minutes and takes 15 minutes, plus a 10 minute drive, so the earliest he will need to leave is 12:25pm.
Hence Ben needs to leave the earliest.

Question 48: B

If the 6 friends all send each other cards, this is a total of 30 cards. 5 of these are to Sophie and 5 are from Sophie. Hence 20 are priced at £0.50, which is a total of £10. A further 5 are priced at £1.50, which is £7.50. The final 5 are priced at £1.20, which is a total of £6.00. Hence the total cost of sending the cards is £23.50.

Question 49: E

In Rovers' first 3 games, they have scored 1 goal and had 8 goals scored against them. In total they scored 1 goal and had 10 goals scored against them, so they must have lost their last game against United 2-0.
In City's first 3 games, they scored 7 goals and had 3 goals scored against them. In total they scored 10 goals and had 4 goals scored against them. Hence they must have won their game against United 3-1. Hence the answer is E.

Question 50: D

Answer E) is not a valid conclusion from the passage, which makes no reference to whether examinations are the fairest method of assessment, simply that they are the fairer than coursework. The passage makes no reference to whether the education minister is logical, so A) is clearly incorrect. B), C) and D) are all valid conclusions from the passage, but we can see that B) and C) contribute towards supporting answer D). Thus, we can see that D) is the main conclusion of this passage, whilst B) and C) are intermediate conclusions, which support this main conclusion.

END OF SECTION

Mock Paper C: Section 2

1. *To what extent are 'logical' and 'rational' synonymous?*

Introduction:

- Consider a definition or overview of each of the terms; rational meaning behaviour or arguments, which are in accordance with reason; logical being in accordance with the rules of formal argument to produce sound inferences.
- Perhaps give some examples of logical and rational situations that illustrate your definitions.
- Introduce a summary of your arguments both for and against the two being synonymous.

Arguments for synonymous:

- Both words are often used interchangeably – a sound explanation can be described as 'logical' and 'rational'.
- We need to be rational to be able to reason logically – logic requires rationality.
- When looking for sensible solutions to situations, adopting a strategy of logic is often an effective way to problem solve and process information.
- Behaviour is often described to be irrational if it doesn't appear to be a logical response to a situation e.g. the logical place to go if you have no money in your purse is the bank, not a shop.
- Irrational behaviour often occurs when people are unable to take time to think through or process situations (e.g. due to emotion or time pressure), and thus do not implement logical reasoning.
- Logic is an effective way of modelling decision making in computer programming.
- Consider the consequences of a world without logic – chaotic, no structure, inability to make the correct decisions in situations.

Arguments against synonymous:

- Rational decisions can be made without fully processing every situation and considering the logical inferences based on each premise.
- Just because behaviour is illogical, does it mean it is irrational?
- Rational responses to certain situations may consider other factors such as emotion, the influences on other people, long term consequences. e.g. it is totally rational to want to avoid a situation that may be the logical decision to make if it will cause personal distress.
- Logic may be too simple to explain all situations.
- Some situations may be unable to be modelled logically or there may be no one logical outcome, yet rationality is still possible.
- Just because there is no logic without rationality, doesn't mean there is no rationality without logic.
- Examples of situations where logic fails or when a logical solution is not possible.

Conclusion:

- Include a summary of arguments both for and against synonymy and draw together both sides of the argument.
- Come to some form of a conclusion – probably a middle ground between logic being one, but not the only, possible way of modelling rationality.
- Emphasise the limitations of simplifying rationality to only logical explanations.

2. *In what instances is aggression justified?*

Introduction:

- Include a summary or definition of aggression – behaviour implemented with the goal of instigating harm to other people.
- Give examples of different types of violence and indicate what types of violence you will be considering in your essay – physical, emotional.
- You may wish to make it clear from the beginning where you stand with regards to the argument – whether it is ever justified, in certain situations or you may wish to suggest here that it is never justified.
- However the essay is also looking for a balanced argument – so summarise both sides, suggestions of when it may be considered justified, but also arguments why this may not be the case.

Possible examples of justified situations:

- When protecting oneself from a physical attack.
- When protecting a loved one from a physical attack.
- The same as above, but verbally- you may wish to consider these two scenarios separately.
- When standing up for a cause e.g. violent campaigns.
- War – fighting for your country.
- In order to control a criminal e.g. police violence towards criminals (may be linked to protection).
- When the target is not another living thing e.g. at the gym, or to let out frustration towards another person on an inanimate object.
- When the target agrees to/or asks for aggression to be displayed – e.g. 'hit me if I do…'
- In jest, comedy, acting e.g. play-fighting.
- In sport e.g. boxing, rugby.

Possible reasons against justification:

- We should never allow aggression in any situation since it makes it more socially acceptable as a behaviour in other, non justified situations.
- Aggressive responses are often learnt from other people therefore we should not model them, in any situation even sport or comedy.
- There are always other ways of letting out frustration, which may be more effective or adaptive, than being aggressive.
- Even if the immediate consequence is not negative, long-term implications of exposure to aggression may not be positive.
- Aggression may also be harmful to the aggressor, as well as the target.
- Any form of aggression has the potential to escalate to cause danger.
- You should never 'fight fire with fire' - Aggression by one person encourages aggression in another.
- Religious/ethical reasons e.g. 'turn the other cheek'.

Conclusion:

- Summarise reasons when aggression is generally not justified.
- Counteract these with a few examples or generalised situations.
- Argue against this to some extent to summarise the other side of the argument.
- You may wish to reach a conclusion about whether aggression is ever justified, and if so, a generalised description of when.

3. *What are the limitations of scientific theories of human behaviour?*
Introduction:

- Explain what a scientific theory is - a generalised, well substantiated explanation of situations or natural occurrences, used as a model and to predict future events or occurrences
- Explain how this may be achieved – 'scientific method' usually requires observable, experimental, objective evidence, repeatedly tested and experimented across multiple situations, beginning with a hypothesis which is proved or disproved through study.
- Make a reference or application of this to 'human behaviour' – explanations or models for what can be observed in individuals emotionally and physically in particular situations.

Paragraph 1: Positive aspects (or in introduction)

Give a brief, general overview of the advantages of scientific theory – this should not be the main focus of the essay, but you may wish to allude to why theories that are 'scientific' are often preferred for example:

- Well tested
- Ability to make predictions and apply to other situations
- Indicate possibilities for further research
- Practical applications e.g. using theories to guide treatments for mental illnesses, guiding business strategies, advertising, etc.

Paragraph 2: Limitations:

You should select a few limitations and expand on each in detail, with examples. Alternatively, select more limitations in less detail, but be careful not to just list the limitations and give a brief description at least of each. Points to consider

- GENERALISABILITY is a major problem – can we generalise human behaviour?
- Reductionist/ too simple: It may not be possible to create a unified model of all situations – there are generally far too many factors involved to accurately predict one situation.
- It may not be possible to even identify all factors contributing to a behaviour– conscious and unconscious influences.
- Individual differences in behaviour.
- Should we be focussing on the differences, rather than the similarities, between human behaviour? How far can generalised accounts take us?
- Ethical considerations e.g. free will – people don't like their behaviour to be able to be predicted by a scientific 'formula'.
- It may be dangerous to emphasise particular aspects of behaviour – may focus wrongly on the wrong aspects, causing certain important aspects to be neglected.
- Is it necessary to try and understand behaviour?
- Limitations of our mind/understanding - Is it possible to understand behaviour or are we aiming towards something unachievable? Will we ever be able to understand ourselves?
- Limitations of scientific research – e.g. practical, not all hypotheseses can be tested experimentally, external validity of the findings.

Conclusion:

- This essay question is so broad that really, you can consider any limitations you wish, so long as sufficient argument/ examples/ explanations are given to back up your point.
- The conclusion should briefly summarise the benefits and limitations of scientific theories.
- You may wish to reach a conclusion as the extent to which scientific models of behaviour should be used.

4. _Assuming time travel was possible, could we learn more from the past or the future?_

Introduction:

- Briefly consider what we might <u>want</u> to learn from looking at the past or future e.g. technology, political decisions, evolution etc.
- Introduce a summary of arguments for both sides of the debate – the benefits of learning from the past, versus the benefits of learning from the future.

Benefits of looking at the past:

- Preventing past mistakes being repeated in the present.
- Understanding previously worked strategies can be used to adapt and produce new strategies to current situations.
- Learning from the knowledge of others requires other people to have experienced things before us.
- History will never repeat itself, but the future we will eventually reach.
- Looking at the past gives us the ability to change the future, but looking at the future does not provide an opportunity to change the past.
- Without an understanding of the foundations/principles of where things came from, we will never be able to fully understand the way things are today, or improve on them.
- A lot can be learned from the way that people lived without the technology, resources, accessibility that we have today.
- Looking at the past makes us appreciate what we have now a lot more.

Benefits of looking at the future:

- If we know what will happen in the future and its demands, we can take steps to be prepared for it e.g. climate change, or change it.
- Again, learning from future mistakes may allows us to prevent them being made so that we can adapt more effectively, knowing what will work and what won't.
- In the future, technology and understanding will advance – therefore if we can look to the future, we can find this out now.
- Focussing on the past leads to regret and guilt, while focussing on the future relieves anxiety and provides an opportunity for things to be changed before they happen – looking at the past can't stop what happened.
- There are endless possibilities in the past, but the future is restricted.
- You may wish to consider ethical implications – briefly consider _if_ we should look into the future - how far would this take us?

Conclusions:

- Summarise both sides of the argument – the benefits of looking at the past and of looking at the future.
- You may wish to consider ethical/philosophical debates about looking into the future.
- Arguably, the future becomes the past if you travel to it and then back – so are we ever learning from the future?
- You should come to a final conclusion/summary about what is the most adaptive strategy, if it were possible, since the question asks for a decision to be made.

END OF PAPER

Mock Paper D: Section 1

Question 1: C

Answer B) is completely irrelevant to what the manager is saying, so is incorrect. A) and E) are also incorrect as the manager is simply talking about ticket sales. He has not mentioned anything about the relevant popularity of folk music, or how much the band should value profit. D) is incorrect as the manager is simply saying that the band will have higher ticket sales in France than in Germany, so other countries are not relevant.

C) is correct as Germany could still have higher ticket sales for folk music than France despite the recent changes in ticket sales.

Question 2: D

E) is completely irrelevant to Rob's conclusion. C) is also irrelevant as Rob has only concluded that Tom is unable to take part, and has said nothing about his own eligibility. B) is incorrect as whether Tom intended to take part in the mountain climbing activities does not affect whether he is able to under the clause Rob has found.

Meanwhile, A) is explicitly stated in the question, as the clause applies to holidays organised by Wild Africa Inc. and Tom and Rob's holiday is organised by Wild Africa Inc. Therefore it is not an assumption.

D) is correct as at no point has it been stated that Tom suffers from *severe* asthma. It has only been stated that Tom suffers from asthma, nothing has been said about the severity. If Tom does not suffer from severe asthma, then Rob's conclusion is incorrect.

Question 3: B

Only B) is not an assumption, as it is stated in the question that both Grace and Rose departed at 5:15pm. The other answers are all assumptions. At no point has it been stated that both the girls are walking, or that they will walk at the same speed. If either of these points are incorrect, we cannot definitely state that they will arrive home at the same time. Therefore A) and E) are assumptions. Also, it has not been stated that the gymnastics class is being held at the local gymnasium. If this is not the case, then we cannot know how far Grace and Rose have to walk, and therefore cannot state that they will arrive home at the same time. Therefore, C) is an assumption. Equally, if Grace gets lost, she may arrive home after Rose, so D) is an assumption.

Question 4: E

Answer A) is completely irrelevant to John's conclusions, as the speed of travel has no effect on the train's destination. D) is also irrelevant as other destinations from King's Cross station also bear no effect on John's conclusion. Meanwhile, B) is incorrect as John's conclusions refer to travelling to Edinburgh by train, so the possibility of travelling by aeroplane has no effect. C) is not an assumption because John's conclusion is in the present tense, referring to journeys made at the moment, so future developments have no effect.

E) is an assumption John has made. Only two other stations in London have been mentioned. At no point has it been mentioned that there are no other stations in London that John could travel from.

Question 5: A

B), C) and D) are all stated in the question, and so are not assumptions. D) is stated as the question states that safety features cause people to drive faster, and then states that higher driving speeds are responsible for many high-speed crashes. E) directly contradicts something stated in the question (the question actually states that high-speed crashes are causing thousands of deaths each year). Also, E) bears no influence on the conclusion of the argument. Therefore, E) is not an assumption.

A) is an assumption because it is required to be true for the argument's conclusion to be valid, but at no point is it stated that a spike would cause people to drive slower.

Question 6: C

B) and D) are both stated in the question. A) is also stated as the question states that Tanks were a hugely influential factor in ALL battles in World War 2.

E) is not stated but is not an assumption as it is not required to be true for the argument's conclusion to be valid.

C) However, is required to be true for the conclusion to be valid and yet is never stated in the question, so it is an assumption.

Question 7: B

D) is irrelevant to the argument's conclusion, whilst A) and E) are also irrelevant as the argument does not directly imply either of these things (and even if it did they are irrelevant to the argument's conclusions so are not flaws).

C) is incorrect because the argument states that the Prussian arrival was essential to the British victory, so C) is not an assumption.

B), however, is never stated in the question, but is needed to be true for the argument's conclusion to be valid.

Question 8: B

D) and E) are both entirely irrelevant to waiting times, so are not flaws.

C) is not correct, as the question states that busier ports have longer queuing times. A) is also incorrect as the question states that Bordeux is the busiest port in France, so Calais is definitely less busy than Bordeux. Therefore, Porto cannot be busier than Bordeux but less busy than Calais.

B) is a flaw, as the fact that Bilbao was busiest last year does not necessarily mean it will be the busiest this year.

Question 9: D

A) is irrelevant to the argument's conclusion, and therefore not a flaw. C) is also irrelevant as other problems stemming from inhalers do not change the fact that inhalers could prevent many hospital trips.

B) and E) are incorrect. The argument only states that inhalers could prevent many hospital trips. It does not assume that inhalers could prevent asthma attacks, and does not state that all the hospital trips could be prevented.

D) is a valid flaw with the argument. At no point is it stated that inhalers stocked in school could be quickly available to children, and yet the argument states that hospital trips could be prevented if inhalers were *quickly available*.

Question 10: A

At no point is A) stated, but if aeroplanes are not a major source of carbon dioxide then it does not follow that they are largely responsible for the damage caused by global warming. Therefore A) is merely an assumption.

B) and C) are both stated in the question, whilst D) is irrelevant to the conclusion. E), meanwhile, is stated, as the question states that *we must now seek to curb air traffic in order to save the world's remaining natural environments*.

Question 11: B

A) and C) can be inferred, as the question states that these things would happen. Meanwhile, D) and E) actually serve to reinforce the argument's conclusion that the research into a new cure will not be successful. Therefore, they are not flaws in the argument's reasoning.

The point raised by B) does weaken the argument, and is a valid flaw in the argument's reasoning.

Question 12: E

A), C) and D) are all irrelevant to the argument's main conclusion, namely that Egypt was a powerful nation and must therefore have had a very strong military.

B) is a conclusion from the argument, but goes on to support E). If a nation required a very strong military to be a powerful nation, then it follows that if Egypt was a powerful nation it must have had a very strong military. Therefore, B) is an intermediate conclusion within the argument. E) is the *main* conclusion of the argument.

Question 13: D

At no point does the argument state or imply that we should not be concerned about damage to the polar ice caps, or that reducing energy consumption will not reduce CO2 emissions. Therefore, B) and E) are incorrect.

C) could be described as an assumption made in the argument, and is therefore not a conclusion.

A) goes beyond what the argument says. The argument does not say there are no environmental benefits to reducing energy consumption; it merely says it will not help the polar ice caps. Therefore A) is incorrect and C) is a valid conclusion from the argument.

Question 14: C

A), B) and D) are all in direct contradiction to statements made in the passage, so cannot be conclusions. E), meanwhile, does not contradict the argument, but at no point does the argument say that the dangerous isotope was not effective at relieving nausea, so E) is not a conclusion.

However, the fact that the company followed the required level of testing and still did not detect the dangerous isotope does suggest that the required level of testing was not sufficient to identify Isotopes, so C) is correct.

Question 15: D

A) and E) are both intermediate conclusions which go on to support the main conclusion, which is that given in D).

B) is irrelevant, and is also not stated or implied in the argument, whilst C) is an assumption made in the argument, and is therefore not a conclusion.

Question 16: A

At no point is it stated or implied that car companies should prioritise profits over the environment, so C) is incorrect. Neither is it stated that the public do not care about helping the environment, so E) is incorrect.

B) is a reason given in the argument, whilst D) is impossible if we accept the argument's reasons as true, so neither of these are conclusions.

The conclusion is that given in A).

Question 17: B

E) is contradictory to the main conclusion of the argument.

A), C) and D) are all reasons which go on to support the main conclusion of the argument, which is given in B). If we accept A), C) and D) as true, then it follows readily that the statement given in B) is true. Therefore, B) is the main conclusion.

Question 18: E

A) is contradictory to the researcher's argument, whilst D) is a completely irrelevant suggestion.

The researcher makes no suggestion that a decrease in productivity will result from not organising a staff day out, so C) is incorrect.

The researcher accepts that there will be a short-term decrease in productivity from organising a staff day out, but that it will be made up for by the long-term increase in productivity. Therefore B) is incorrect, as the researcher is not suggesting the short term decrease is ignores. E) is more in line with the researcher's argument, so E) is the main conclusion.

Question 19: C

B) and D) are irrelevant suggestions that can be safely ignored. E) is in direct contradiction to a statement in the passage that the 2 candidates were clearly more suitable than the others.

A) is contradictory to what the manager has said, whilst C) follows on from what the manager says. Therefore, C) is a conclusion which is much more readily drawn from the passage.

Question 20: E

The passage is suggesting a system should be used in which people pay for *whichever* services they use. The new system is not referring to *how much* a customer uses a service, simply *which* services are used.

All of the answers follow this reasoning except for E). E) refers to a system where people pay according to *how much* they use a service, rather than paying for whichever services they use. Therefore, E) follows a different line of reasoning from the passage.

Question 21: B

B) is an underlying assumption in the Transport Minister's argument. If rural areas have plenty of passengers, her assertion that rail companies will not run many services to these areas does not follow from her reasoning. Therefore, if B) is true, it strengthens the transport minister's argument.

Meanwhile, D) would actually weaken the transport minister's argument, suggesting that privatisation would not lead to less service for rural areas.

C) is irrelevant as the transport minister is arguing about how rural communities will be cut off by a privatised system. She is not referring to the quality or price of rail services under a publically subsidised system.

A) and E) are completely irrelevant points, which have no effect at all on the strength of the Transport Minister's argument.

Question 22: C

The Argument follows a pattern of "**IF** A happens, B **WILL** happen. A Happens, therefore B **WILL** happen". C) is the only answer which follows this pattern of reasoning.

A) reasons as "**IF** A happens, B **WILL** happen", but then incorrectly concludes that "If B is to happen, A **must** happen". This is incorrect. In this example, there may be other cheaper airports to fly from apart from Gatwick.

D) is also incorrect, stating that "A **must** happen for B to happen" but then concluding "A happens, therefore B **will** happen". This is incorrect, as the fact that A is needed for B to happen offers no guarantee that B will happen if A does.

B) and E) both follow correct reasoning, but different to that in the question. B) and E) reason as "A **must** happen for B to happen. B happens, therefore A **must** have happened". This is not the same as saying "**If** A happens, B **will** happen", as in the question.

Question 23: E

The passage follows principle that everybody should pay an amount depending on how wealthy they are, which is then used to fund services that are allocated based on need.

B), C) and E) both refer to services being allocated based on need, not ability to pay. However, C) refers to companies responsible for the problems being made to pay for the service, and B) refers to everybody paying a fixed amount to fund the service. E) refers to people paying an amount dependent on personal wealth. Therefore E) best illustrates the principle of the passage.

The other answers do not refer to a service being allocated based on need. D) directly contradicts this principle, whilst A) is not referring to a service that is being funded/provided.

Question 24: A

B) and E) are irrelevant points which do not affect the strength of Lucy's argument.

C) and D) would both serve to strengthen Lucy's argument. C) suggests that running costs will be low, whilst D) suggests that visitor centres are profitable. Both of these, if true, serve to suggest that opening up visitor centres will be profitable for the park, therefore supporting Lucy's argument.

A), however, would weaken Lucy's argument by suggesting that visitor centres will not be profitable.

Question 25: D

The principle used in the argument is that there is no point in engaging in debate with people who are wrong about the facts on a particular issue. Only D) follows this line of reasoning.

B) and E) are both in direct contradiction to this principle, whilst A) is an irrelevant point, not referring to who we should engage in debate with.

C) refers to who Kerry should talk to, but does not explicitly state that she should not talk to those who are wrong about the issue. Therefore C) is not as effective an illustration of the principle at hand as D).

Question 26: D

If the announcement is accurate to the nearest 10 minutes, this means that the soonest the train will arrive in London is 115 minutes after the announcement, which is 17:25. The final destination is 10 minutes from King's Cross, so the earliest time I might arrive there is 17:35.

Question 27: C

When the sister's son is home, she has to buy a carton of washing powder 1.2 times as often, so she must be doing 1.2 times as many loads of washing. If we let x be the number of people living at home when the son is home, the number of loads of washing when he is home is 2+x, whereas the number when is not at home is 2+x-1=1+x. Therefore 2+x must equal 1.2(1+x). Rearranging this we get 0.8=0.2x, so x is 0.8 divided by 0.2, which is 4. So 4 people are living in the house when the son is home.

Question 28: D

The total time each train can run is 18 hours a day. Each journey takes the train 30 minutes (24+4+2). So each train can make 36 journeys a day. Therefore the total journeys made by the shuttle service per day will be 2x36 (because there are 2 trains) so the answer is 72.

Question 29: E

The lowest common multiple of 6, 4 and 2.5 is 60. Hence trains from all 3 lines will arrive at the same time every 60 minutes. If the last time they did was 4 minutes ago, it will hence be 60-4=56 minutes until they do so again. Therefore the answer is E.

Question 30: E

If Sam buys the invitations, she will spend £90 (90 x £1 each).

If Sam makes the invitations, she will need enough supplies for 94 invitations, which will be:

➤ 4 packs of red paper at £2 each = £8
➤ 7 rolls of ribbon at £3 each = £21
➤ 4 packs of gold stickers at £1 each = £4
➤ 1 stamper = £8
➤ 2 ink pads at £4 each = £8
➤ 5 packs of cream card at £2 each = £10

Adding these up, we get that the total spent on making the invitations herself is £59. Therefore she saves £90-£59=£31 by making them rather than buying them. Therefore the answer is E.

Question 31: B

If there are at least as many boys as girls in the class of 36, then there are at least 18 boys and at least 9 boys have brown eyes. If two thirds of the class have brown hair and at least as many boys as girls have brown hair, at least two thirds of the boys in the class have brown hair, so at least 12 boys have brown hair. There are only 18 boys in the class, so of the 9 boys who have brown eyes and 12 who have brown hair, at least 3 of these must be the same boys. So at least 3 boys in the class have both brown hair and brown eyes.

Question 32: C

The total amount of dilute squash needed is (300ml x 8) + (400ml x 3) = 3600ml. Mandy has 600ml of concentrated squash so she needs 3000ml of water to make up the right amount. There should be 3000ml:600ml of water:concentrated squash so the ratio needed is 5:1

Question 33: D

If we assumed that none of the options in any block were the same, there would be 4x4x4x4 = 256 different sets of options. However, Psychology and Mathematics are in 2 different blocks so there will be some options that cannot be taken together. There are 16 sets of 4 options that involve Mathematics twice (Mathematics in Blocks A and C and then any combination of options in the other blocks) and 16 that involve Psychology twice (Psychology in Blocks A and B and any combination of options in the other blocks), so we need to take off 32 from the total options we calculated. Hence the total number of sets of possible options is 256-32=224.

Question 34: C

We need to work out both what the cost of living will be in 4 years time and how much savings there will be in 4 years time.

The cost of living rises by 3% a year, so using the table we can see that at the end of Year 1 the cost of living is £19,570.00. Then at the end of Year 2, the cost of living is £20,157.10. At the end of Year 3, the cost of living is £20,761.81. At the end of Year 4, the cost of living is £21,384.67.

The savings rise by 4% each year until they exceed £20,000, so after Year 1 they are £19,760.00. After Year 2, they are £20,550.40. This means that for the last 2 years, they will yield 5%. So after Year 3, they will be £21,577.92, and after Year 4, they will be £22,656.82.

We can then work out the difference between these two amounts. We only need an approximation, so we can do £22,656 - £21,384 = £1272, which to the nearest £100 is £1300. Hence the answer is C.

Question 35: D

Amaia needs to arrive by 1100 so the latest bus she can catch is the one which arrives at Northtown University at 1053. This arrives at Northtown West station at 1019, or Northtown Central station at 1040. The latest train she can catch to make this bus is the one that gets to Northtown Central for 1028, and leaves Southtown at 0943. Amaia lives a 10 minute walk from Southtown station so she will need to leave her house at 0933 to get to the interview on time. Therefore the answer is D.

Question 36: C

The instructions say that Shaniqua plays in the square which will stop Summer being able to win straight away, so Shaniqua must play in 4. Summer then needs to play in a square where there will be 2 different options to make a line on the turn afterwards, so that Shaniqua cannot block both of them. If Summer plays in 1, she can make a line by playing in either 5 or 6 the next turn, so Shaniqua cannot stop her winning. If Summer plays in 2, she cannot make a line on the next turn at all. If Summer plays in 3, she can only make a line by playing in 6 the next turn and so Shaniqua can stop her. If Summer plays in 5, she can only make a line by playing in 5 the next turn and so Shaniqua can stop her. If Summer plays in 6, she can make a line by playing in either 1 or 3 the next turn, so Shaniqua cannot stop her winning. Therefore she either needs to play in 1 or 6 to be able to be certain of winning the next time. Therefore the answer is C.

Question 37: E

There are 16 squares of dimension 1. There are 9 squares of dimension 2 (one in each corner, one halfway across each side and one right in the middle). There are 4 squares of dimension 3 (one in each corner). There is 1 square of dimension 4. Therefore the total number of squares is 16+9+4+1=30. Therefore the answer is E.

Question 38: D

Any date in October 2014 will have a 1 in both the month and year so we have to go back to September. 30th September will have a 0 in both the date and month. 29th September will have a 9 in both the date and the month. Therefore the last date with every number different in the 6 digit format will be 28-09-2014, which is 14 days before the 12th October. Therefore the answer is D.

Question 39: E

Ashley has to be sat in the front left seat so there are only two seats left in the front row. Bella and Caitlin have to be sat in different rows, so one of them must be sat in the front row and one in the back row. Now there is only one seat left in the front row, so there is not room for Danielle and her teaching assistant to both sit there. Therefore Danielle and the teaching assistant must take the two remaining seats in the back row. Therefore Emily must sit on the front row as there are no seats remaining in the back row. Emily cannot sit in the middle seat due to her mobility issues, so she must sit in the front right seat. Therefore the answer is E.

Question 40: B

➤ The volume of the box with 10cm squares cut out is $10*100*100 = 100000cm^3$
➤ The volume of the box with 20cm squares cut out is $20*80*80 = 128000cm^3$
➤ The volume of the box with 30cm squares cut out is $30*60*60 = 108000cm^3$
➤ The volume of the box with 40cm squares cut out is $40*40*40 = 64000cm^3$
➤ The volume of the box with 50cm squares cut out is $50*20*20 = 20000cm^3$

Therefore the biggest box is the one with the 20cm squares cut out, so the answer is B.

Question 41: E

We can calculate all the rental yields as follows:
➤ House A: $(700x12)/168000 = 0.05$
➤ House B: $(40x125x4)/200000 = 20000/200000 = 0.10$
➤ House C: $(600x12)/144000 = 7200/144000 = 0.05$
➤ House D: $(2000x12)/240000 = 24000/240000 = 0.10$

House E: $(200x52)/100000 = 10400/100000$. We can see by observation that this is > 0.1 as $10000/100000$ would equal 0.1, therefore there is no need to work this out to be able to say that this is the house with the highest yield.

Question 42: D

Let the number of invitations with the extra information in be m. Invitations with extra information in cost £0.70 to send and invitations without cost £0.60. Therefore the total cost of posting is £0.70m + £0.60(50-m) and this is equal to £33. 33=0.70m-0.60m+30. 3=0.1m therefore m=30. So the number of invitations with extra information in is 30. Therefore the answer is D.

Question 43: D

If the median is 6, the 3rd number when the numbers are written in order is 6. If the mode is 4, there must be at least two 4s (and can only be two 4s, because there are only 2 numbers less than 6 due to what we know about the median). Therefore the smallest three numbers in the set are 4, 4, 6. For the mean to be 8, the numbers must add up to 5 times 8 = 40. Therefore the largest two numbers must add up to 40-(4+4+6)=26. Therefore the answer is D.

Question 44: D

The hospital director's comments make it abundantly clear that the most important aspect of the new candidate is good surgery skills, because the hospital's surgery success record requires improvement. If we accept his reasoning as being true, then it is clear that the candidate who is most proficient at surgery should be hired, and patient interaction should not be the deciding factor. Thus, Candidate 3 should be hired, as suggested by D.

Question 45: B

We can calculate the prices if the child's fare is a half of the adult fare, then as if the child's fare is three quarters of the adult fare, and choose the multiple of 20 that lies between the two.

If the child's fare is half the adult fare, then £4.20 is twice the adult fare as this is the price for one adult and two children each of whom pay half the adult price, so it is the equivalent of paying for 2 adults. Hence the adult fare is £2.10 and the child fare is £1.05.

If the child's fare is three quarters of the adult fare, then £4.20 is two and a half times the adult fare. Hence the adult fare is £1.68 and the child fare is 84p.

So we know the child fare is between 84p and £1.05. We also know it is a multiple of 20p. Hence the child fare is £1. The adult fare is between £1.68 and £2.10, so on the basis that it is a multiple of 20p it could either be £1.80 or £2. However, we know the child fare is more than half the adult fare, so it cannot be £2 as this would mean the child fare was half of the adult fare. Therefore the adult fare must be £1.80 and the difference between the two is £1.80-£1 = 80p. Therefore the answer is B.

Question 46: A

Each group involves 10 matches (each team plays each other team once, so if the teams are A through to E, the 10 matches are A-B, A-C, A-D, A-E, B-C, B-D, B-E, C-D, C-E, D-E). There are 8 groups, so in total in the group stage there are 80 matches. 16 teams then progress from the group stage, meaning 8 matches in this round. There are then 4 matches in the next round, 2 matches in the round after and then is the final. Therefore in total there are 80+8+4+2+1 matches, = 95.Therefore the answer is A.

Question 47: E

We can work out the code for each number and see which one equals 3.
➢ The code for A is (3x4) = 12, divided by 6 = 2, minus 1 = 1
➢ The code for B is (9x8) = 72, divided by 6 = 12, minus 4 = 8
➢ The code for C is (5x4) = 20, divided by 2 = 10, minus 3 = 7
➢ The code for D is (7x8) = 56, divided by 4 = 14, minus 8 = 6
➢ The code for E is (6x8) = 68, divided by 4 = 17, minus 9 = 3
Therefore the pin number with the code 3 is E, 6849.

Question 48: C

If it will cost Niall £2 more to pay per session than to buy membership, the combination of classes and gym sessions he is going to attend must cost £32. The only one of these combinations which costs £32 is C (5 x gym sessions at £4, 6 x classes at £2). Hence the answer is C.

Question 49: E
If I can lift boxes up to half my weight, the total I can lift at a time is 30kg. If the empty box weighs 0.5kg and each book is 2kg, up to 14 books can be lifted per box (so each full box weighs 28.5kg). 120 divided by 14 is between 8 and 9. Hence it will take 9 trips to move all the books.

Question 50: E
- At the end of the first year, Jenny gets a rise of 3%, which is £720, rounded up to the nearest £1000, which is £1000. Hence her salary at the end of the first year is £25000.
- At the end of the second year, Jenny gets a rise of 6%, which is £1500, rounded up to the nearest £1000, which is £2000. Hence her salary at the end of the second year is £27000.
- At the end of the third year, Jenny gets a rise of 9%, which is £2430, rounded up to the nearest £1000, which is £3000. Hence her salary at the end of the third year is £30000.
- At the end of the fourth year, Jenny gets a rise of 12%, which is £3600, rounded up to the nearest £1000, which is £4000. Hence her salary at the end of the fourth year is £34000.
- At the end of the fifth year, Jenny gets a rise of 15%, which is £5100, rounded up to the nearest £1000, which is £6000. Hence her salary at the end of the fifth year (and hence after 5 and a half years) is £40000. Hence the answer is E.

END OF SECTION

Mock Paper D: Section 2

1. *Describe a scientific model of a human brain. Justify any materials you might need, how you would connect the separate components and highlight any limitations your model might have.*

This is a very broad question and can be taken in many different ways – there is no correct model. The essay should be structured as follows:

- Introduction – general overview of the model, describing the important elements of the human brain and how they could be portrayed into a model. The introduction may mention the limitations of any model of a brain, in that we do not understand the human brain fully and it is far too complex to be modelled by anything other than itself.
- Main body – organized into paragraphs describing each component. All models should follow a logical structure, with clear justification for each decision made. There should be a continual reference to the human brain in the justification and clear links between paragraphs describing how the model could be achieved.
- Conclusion – this should summarize the main components of the model. Here it would be relevant to mention any major limitations, or any possible developments that could be made to the model in the future. The model need not be complete, and can be interpreted however the student wishes, so long as there is justification for each decision.

An example of a model of the human brain could be:

- A series of computers, each programmable to different functions of the human brain, joined together with a series of wires acting as the neuronal connections between each component.
- Limitations of a system such as this could be size, as the number of computers required would be too numerous to describe, and not at all proportional to the human brain. We are also limited by our human understanding, both of technology and computing, and of the brain itself. Brains are adaptable and have 'plasticity' – a computerized model would not be able to show this feature, since they will have to be programmed in advance of their function. In this way, any model of a human brain would be reductionist.

Any other model of a brain is credible and equally successful so long as it is logical and key components are described and justified in detail.

2. *"Impatience is the driving force behind development." To what extent is patience no longer a virtue in modern society?*

Introduction:

- Consider a definition of patience – 'the capacity to accept or tolerate delay, problems, or suffering without becoming annoyed or anxious' and impatience as the opposite to this.
- You may wish to explore the phrase 'patience is a virtue' – where a virtue is a character trait which is considered good or desirable.
- Make a reference to the original quote – which suggests that it may be impatience that is valued with the way that society is rapidly changing and developing. Emphasis on 'modern' society – is the phrase outdated?
- Introduce ideas on both sides of the argument.

Possible arguments for:

- Arguable that to be successful, you must be impatient for change and success – if you aren't, you aren't passionate enough and you won't succeed. Good leaders are driven and motivated for change, which often comes from an impatience.
- The speed that technology (for example) is advancing and changing means that you have to be impatient for the next thing otherwise you will always be one step behind.
- Learning is also helped by impatience – the more you want to know NOW, the harder you will work to understand it. Impatience at your current capabilities and understanding will lead to a drive a passion to learn and comprehend more.
- Impatience is perpetual – the quicker we get what we want, the quicker we want it next time. Therefore impatience is required to stay driven and committed.

Possible arguments against:

- Impatience leads to stress, which can have several impacts – on health, on relationships, on work. Therefore the consequences of impatience can be detrimental and it is undesirable to be impatient.
- Patience leads to calmness and relaxation, which are important for the ability to rest our bodies and taking time to enjoy the world. If we forget how to be patient, we will stop having time for ourselves, and, even worse, time for other people.
- We have no need to be impatient in such a fast paced society – things are changing so quickly, we can access what we want so quickly with technology that impatience is unnecessary and patience should be practiced before we forget how to employ it.
- Patience allows us to value the world – it gives us time to think, and stops us taking everything for granted.
- Patience is a character trait which is looked for in friendships – everyone wants a friend who they feel will listen and will spend time with them, not someone who is constantly in a rush for the next thing to happen.

Conclusion:

- Summarize the main points made on each side of the argument in the essay.
- Link back to the original quote – mention development and drive.
- You may wish to come to a decision either way, or it is equally fine to sit somewhere in the middle, so long as this is fairly justified.

3.   *Voltaire said, "The perfect is the enemy of the good." Discuss this claim.*

Introduction:

- Consider a definition of perfect - having all the required or desirable elements, qualities, or characteristics; as good as it is possible to be, versus good - to be desired or approved of, of a high standard.
- You may wish to discuss the term 'enemy' – by this it something that is directly opposed to something else.
- Best to consider this in the context of striving for each quality – is either ever achievable?
- Introduce the points you will make on both sides of the argument.

Possible arguments for:

- Perfection is unachievable in a world that has so much wrong and badness in it – therefore striving to be perfect is impossible and will only lead to disappointment.
- Goodness cannot be appreciated fully if the ultimate goal is perfection, because everything below that goal will not be good enough. Therefore a state of perfection eliminates the possibility of goodness because nothing, other than perfection, is 'good enough'.
- Striving for perfection causes inevitable failure, and failure is not desirable, and (since good is defined as something to be desired), therefore it is not good.
- Example of clinical perfectionists – unhappy, constantly working, can lead to or be connected with mental health issues.

Possible arguments against:

- Perfection and goodness go hand in hand, while striving for goodness, you are getting closer and closer to perfection. Therefore perfection aids goodness.
- Things can only look good in relation to the superlative – so how could we judge goodness without perfection?
- Consider a sliding scale between perfect and imperfect – good is closer to perfect, and therefore cannot be considered its enemy.
- If perfection is unachievable, but goodness is, you can find goodness in the pursuit of perfection, so long as you allow for something to be 'good enough'.
- Perfection is 'ultimate goodness', but goodness is only 'not good enough' if we decide it to be so – it is all a matter of perspective and goal setting.

Conclusion:

- Summarize the main points on both sides of the argument.
- Link back to the original quote – you may wish to bring up the idea of friends and enemies again.
- You may choose to ultimately agree or disagree with the claim, or sit in the middle with justification on either side – so long as the argument is logical following from the rest of the paragraph, either is equally credible.

4.   *Is world peace achievable?*

Introduction:

- Consider a definition of peace- freedom from disturbance, tranquillity, the absence of conflict. Or relate this directly to world peace - freedom, cooperation and happiness among and within all nations and people.
- It may be relevant to mention the number of wars and the amount of conflict going on around us at the moment – both in wars and more silently, in areas of corruption and poverty.
- You could introduce the ways in which it could be reached – either voluntarily or by virtue of a system of governance that prevents warfare. Is the latter true peace?
- Introduce the ideas on both sides of the argument you wish to discuss.

Possible arguments for:

- Achievable is a goal that is possible – and it is certainly possible for all nations to cooperate, given the correct guidance and when there is will on all sides
- The first step is believing it is possible, and we do not know the future. An 18th century European would never believe that German, French, Austrians etc could ever live peacefully together, but this has happened. When there is constant change, we cannot decide that anything is impossible. When there is a will, there is a way.
- So long as there is a dream and the concept is still alive, there is always the possibility that all nations can live happily together. Cooperation can take place, people can unite.
- The world is made up of people, each with their own thoughts and their own beliefs. We are all human, we all make mistakes, but there is also good in everybody and when striving for a common goal, that can be reached.
- The importance should be placed on a strategy and an ultimate goal – a system working together, involving individual people joining as one, such as in government, can always work for peace. We have the quality within us so long as we choose to use it.

Possible arguments against:

- It is easy to get carried away with dreaming – just because something is possible doesn't always mean it is achievable? It is not realistic to think that we can solve such a massive issue.
- There are too many variables involved – religion, economics, race, etc and too many opinions. How could we achieve everyone seeing eye to eye when everyone has different opinions?
- Look at history – as one war ends, another one begins. We learn from the past, and the past has shown that humans cannot live harmoniously with each other.
- As long as there is badness in the world – sin and corruption- in the form of greed, impatience, anger, selfishness – there will always be conflict. And world peace would be achieved if there were no conflict, so it is not achievable.

Conclusion:

- Summarize points made on both sides of the argument concisely and draw ends together.
- Link back to the original point of world peace as being an ideal, and ultimatum – achievable meaning something can be reached as a goal. Ensure all arguments are relevant to this point.
- It is credible to either come to a conclusion for or against world peace, or to come to a compromise – such as world peace being possible as a goal, but with conditions or some consideration for conflict or argument.

END OF PAPER

Mock Paper E: Section 1

Question 1: C
The simplest solution is to calculate the total area at the start as 20 x 20 = 400cm^2. Then recognise that with every fold the area will be reduced by half therefore the area will decrease as follows: 400, 200, 100, 50, 25, 12.5 – requiring a total of 5 folds.

Question 2: D
This is the only correct as it is the only statement that doesn't categorically state a fact that was discussed in conditional tense in the paragraph.

Question 3: B
Off the 50% carrying the parasite 20% are symptomatic. Therefore 0.5 x 0.2 = 10% of the total population are infected and symptomatic. Of which 0.1 x 0.9 = 9% are male.

Question 4: D
The most important part of the question to note is the figure of 30% reduction during sale time. Although A and B are possible the question asks specifically with regard to cost. Therefore, it is only worth waiting for the sale period if the sterling to euro exchange rate does not depreciate more than the magnitude of the sale. As such solution D is the only correct answer as it describes anticipating a loss in sterling value less than 30% against the euro.

Question 5: D
Begin by calculating the number of childminders that can be hired for a 24-hour period as 24 x 8.5 = 204. Therefore, a total of 4 childminders can be hired continually for 24 hours with £184 left over – as the question states the hire has to be for a whole 24-hour period and therefore the remainder £184 cannot be used. As such D is the correct answer of 4 x 4 = 16.

Question 6: B
The simplest way to approach this question is to recognise that there is a difference of £1.50 between peak and off-peak prices for all individuals except students. The total savings can therefore be calculated as (3 + 5 + 1) x 1.5 = 9 x 1.5 = 13.5.

Question 7: A
Karen is a musician, so she must play an instrument, but we do not know how many instruments she plays. Although all oboe players are musicians, it does not mean all musician play the oboe. Similarly, oboes and pianos are instruments, but they are not the only instruments. So, statements b and c are incorrect. Karen is a musician but that merely means that she plays an instrument, we do not know if it is the oboe. So, statement d is incorrect.

Question 8: D
Answers A and B are simply incorrect as the measurement taken is a percentage increase (/decrease) which will normalise baseline diameters therefore allowing for comparison over multiple time points. You should be aware from your studies that ultrasound is an invaluable technique in distinguishing between adjacent tissue types. Any methodology is repeatable if it is correctly chronicled and followed therefore leaving the correct answer of D.

Question 9: C
If both the flight and travel from the airport are delayed this will be the longest the journey could possible take – producing a total journey time of 20 + 15 + 150 + 20 + 25 = 230 minutes or 3 hours 50 minutes. Therefore given all possible eventualities, to arrive at 5pm, boarding should begin at 13.10pm. Answer D is incorrect as a delayed plan would add 20 minutes to the journey whilst the transport to the meeting at the other end takes a minimum of 15 minutes – even if Megan could teleport instantaneously from the airport to the meeting she would be 5 minutes later than if there wasn't a plane delay.

Question 10: E
This is almost a trick question and simply an application of exponential decay. Recall that an exponential decay is asymptotic to 0 as no matter how small the volume within the cask becomes, only half of it is ever removed. It could be argued that this process cannot continue once a single molecule of whiskey is left – and when splitting that single molecule in half it is no longer whiskey. However, the question does not ask "how long till all the whiskey is gone" but rather "how many minutes will it take for the entire cask to be emptied" and therefore the process can continue infinitely – even if the only thing left in the cask is a collection of quarks … or half that.

Question 11: D
A sky view of the arrangement leads to:

C	A	B				A	C	B
D		E		*or*			D	E

In both, D is to the left of E, thus is the only correct answer.

Question 12: E
The question can be expressed as $(40 \times 30) - x(50 \times 30) = 200 = 1,200 - x1,500$. Therefore $x = 2/3$.

Question 13: D
The information given is very much a red herring in this question. This can be solved using your own application of sequence theory. If a sequence is 6 numbers long, the 6 numbers can be re-arranged into a total of 6! possible sequences = $1 \times 2 \times 3 \times 4 \times 5 \times 6 = 720$

Question 14: E
Although this is an extremely abstract question, all the information needed to answer it is provided. The key rule to have isolated from the information is that with each progressive sequence, a bell can only move one position at most. Therefore, looking through each in turn – for example we can exclude A as the 3 goes from position 4 to position 2. This leaves the only correct answer of B. Recall that the information states bells can only move a maximum of one place – they can of course move 0 places.

Question 15: A
As the largest digit on the number pad is 9, even if 9 was pressed for an infinitely long time the entered code would still average out at no larger than 9. Therefore, it would be impossible to achieve a reference number larger than 9. Indeed, this is an extremely insecure safe but not for the reason described in B (for if the same incorrect number was pressed indefinitely it would never average out as the correct one) but rather because the safe could in theory be opened with a single digit.

Question 16: D
A is incorrect as it ignores the section of the text that states the evolution of resistant strains is driven by the presence of antibiotics themselves. The text states that the rate of bacterial reproduction is a large contributing factor and therefore not wholly responsible – hence B is incorrect. Since this is just one example (and only the information in the text should be considered for these questions) for C to make such a general statement is complete unjustified.

Question 17: B
The fastest way to solve this question is to calculate the quantity of cheese per portion as 200/10 = 20. Which for 350 people would require 350 x 20 = 7000g or 7kg.

Question 18: C
Calculate the calorific content of 12 portions as 12 x 300 = 3,600kcal. As this represents 120%, evaluate what the initial amount would be as (3,600/120) x100 = 3,000kcal.

Question 19: B
Begin by calculating the initial weight of all the ingredients in the Bolognese sauce which comes to a total of 3.05kg. Therefore when cooking for 10 people 3.05 x 4 = 12.2kg of pasta should be used. Which in turn means for 30 people 3 x 12.2 = 36.6kg should be used.

Question 20: D
Calculate the new weight of ingredients in the Bolognese sauce excluding garlic and pancetta which produces a total of 2.8kg. Note that onions represent 0.3kg per 10 people and as such the ratio can be represented as 0.3/2.8 or alternatively dividing top and bottom by 0.3 → 1/9.3

Question 21: E
Begin with calculating total preparation time as 25 x 4 = 100 mins. The fact that Simon can only cook 8 portions at a time is somewhat a red herring as it doesn't impact the calculation. Total cooking time can be calculated as a further 25 x 8 = 200 mins. Producing a total time of 300mins or 5 hours.

Question 22: D
Due to the quantities colour t-shirts are priced at £5 and black and white at £2.50. Therefore, the order will incur a total cost of 50 + (50 x 5) + (200 x 2.5) = £800.

Question 23: E
Answers A and B directly conflict with information presented in the text whilst C and D may well be true but there is insufficient information in the text to address these points. This is an important reminder that although you may well have been a scout and be able to comment on options C and D, you may only consider these arguments in terms of the text provided.

Question 24: B
There are two directions: clockwise and anticlockwise and rats will only collide if they pick opposing directions.
Rat A - clockwise, Rat B - clockwise, Rat C - clockwise = 0.5 x 0.5 x 0.5 = 0.125.
Rat A - anticlockwise, Rat B - anticlockwise, Rat C - anticlockwise = 0.5 x 0.5 x 0.5 = 0.125.
0.125 + 0.125 = 0.25. So, the probability they do not collide is 0.25.

Question 25: C
The article seen here is a particularly good effort at a discursive text as it is completely impartial. Note that the article simply states the facts from either side in equal measure. Nowhere does the author present their opinion on the matter nor do they insinuate their beliefs in anyway.

Question 26: C
The journey time is rounded to the nearest hour (13). Therefore, the longest it could possibly be is 13 hours 29 minutes or it would be round up to 8 hours. Therefore, the latest the ferry will arrive, assuming the travel time estimate is accurate, is given as 20.29.

Question 27: B
The correct answer is 28. Assume, although very unlikely, that you roll a 2 every single go – you will never need to take a step back, only your two forward. Therefore, rolling a lower number in this case is beneficial.

Question 28: B
The simplest way to calculate this is to find the lowest common multiple of the given laptops which is 40 x 60 x 70 = 168,000 seconds = 2,800 minutes = 46.67 hours – although they will not all be on the same lap number at this time.

Question 29: C

A large amount of subtly different data is described here. Of note is the first experiment which describes how nerve conduction is faster in right handed men than it is in left handed men. This result is not transferable to women until it is proven! The experiment currently being conducted only considers dominant hand in men vs. women. That could be either hand or not necessarily the females' right hand. For example, all of the females in this experiment could have been left handed and there is no information in the text to say otherwise, therefore we cannot tell.

Question 30: C

Recognise that two square based pyramids will comprise 8 triangles of base width 5 and height 8; plus, two 5 x 5 squares. Thus, giving at total area of $8(8/2 \times 5) + 2(5 \times 5) = 210$ cm^2

Question 31: B

In order to approach this question first realise that in the first well 1ml of solvent is being combined with 9ml of distilled water producing 1eq of solute in 10ml – hence the first well produces a dilution by a factor of 10. With each progressive dilution the concentration is reduced by a further factor of 10 – hence by well 10 the concentration is at $x/10^{10}$

Question 32: E

The compartments of the human body are occupied by numerous fluids, as the student is only interested in measuring the volume of blood, it is essential he chooses a solute that will only dissolve in blood. So as his known quantity of solute remains no, it must be neither removed nor added during its time in the body. Hence all of the written assumptions must be made and many more.

Question 33: D

The fastest way to approach this question is by calculating the total price per head for the cheapest option as $10 + (8/20) + (10/60) = 10.567$. To make the maths simpler this can be rounded safely to £11 a head at this stage. 2,300/11 is approximately equal to 209 which when rounded to the nearest 10 is 210 people.

Question 34: A

Whilst this passage is attempting to weigh up two sides of an argument, it has a clear one-sided approach focussing heavily on the excitement of dangerous sports. It even states that hunting is recognised as exciting by some. Since the previous sentence discussed the link between archery and hunting, the statement is a fair extrapolation to make.

Question 35: A

If society disagree that vaccinations should be compulsory, then they will not fund them. So, statement A is correct. It attacks the conclusion. Statement b - society does not necessarily mean local so this does not address the argument. Statement c strengthens, not weakens, the argument for vaccinations. Statement d – the wants of healthcare workers do not affect whether vaccinations are necessary.

Question 36: C

Start by calculating the area of wall that may be painted per tin of paint as 10 x 5 = 50m^2. Therefore, to paint the whole area 1050/50 = 21 tins of paint are required per coat. As such to complete 3 coats it will cost Josh 3 x 21 x 4.99 = 314.37.

Question 37: C

A is a correct assumption as procession is a function of rotational motion. B is a necessary assumption or rather inference of the first sentence. The second sentence only says that an asterism can be used, not that it is the only possible method. Nothing is mentioned of navigating the Southern Hemisphere and therefore C is not a valid assumption.

Question 38: D

Recognise that "bank hours" refers only to hours that the bank is open – which Mon to Fri is 8 hours whereas it is only 6 hours on a Saturday. Although John needs the money by 8pm the bank closes at 5 and that 3 hours difference cannot be used. Hence working backwards John will need 8 hours on the Tuesday, 8 hours on the Monday, Sunday is closed, 6 hours on the Saturday, 8 hours on Friday and 8 hours on Thursday and 4 hours on the Wednesday. With a closing time of 5pm, the latest John can cash the cheque on Wednesday is 1pm.

Question 39: D

First thing to recognise here of course is that individual diamonds can be combined to form larger diamonds with the 5 x 5 diamond the biggest of them all. To avoid counting them all and risking losing count, instead deduced the number of triangles per corner and per side; then multiply up by 4.

Question 40: B

Let my current age = m and my brother's current age = g. The first section of this question can therefore be expressed as $m + 4 = 1/3(g + 1)$ whereas the second half can be represented as $2(m + 20) = g + 20$. Therefore, this problem can be solved as simultaneous equations. Rearranged the second equation reads $m = 1/2g - 10$; when substituted into the first equation we form $1/2g – 10 + 4 = 1/3(g + 1)$. Expand and simplify to $1/2g – 6 = 1/3g + 1/3$ → $1/6g = 6\frac{1}{3}$ which therefore means my brother's current age $= 6\frac{1}{3} / (1/6) = 114/3 = 38$. Which means that my current age $= 1/2(38) – 10 = 9$.

Question 41: D

A is categorically wrong as the first two paragraphs discuss how aneurysms produce inflammation which in turn blunts endothelial NO action. B is incorrect as it states aneurysms directly promote CVD, this is not a direct process. It is the blunted NO which directly produces the CVD. C can be ignored as nowhere are aneurysms categorised like this. E is incorrect as the text states that aneurysms reduce NO which will reduce vasodilatation, thus increasing basal vasoconstriction and thus reducing blood flow. Leaving the correct answer of D which is of course true as observations are not transferable between species until tested scientifically.

Question 42: C

Any statement which refers to national or global figures is instantly incorrect as the text does not mention any statistical analysis has taken place. In order to produce national statistics from a small sample size such as this requires statistical analysis. Whilst E could possibly be true it cannot be stated as there are so many possibilities – perhaps the time of the survey was during rush hour in which case the majority of the traffic would have been travelling in the same direction anyway to reach an industrialised area.

Question 43: B

The runners aren't apart at a constant distance; they get further apart as they run. Xavier and Yolanda are less than 20m apart at the time William finishes. Each runner beats the next runner by the same distance, so they must have the same difference between speeds. When William finishes at 100m and Xavier is at 80m. When Xavier crosses the finish line then Yolanda is at 80m. We need to know where Yolanda is when Xavier is at 80m. William's speed = distance/time = 100/T. Xavier's speed = 80/T. So, Xavier has 80% of William's speed. This makes Yolanda's speed 80% of Xavier's and 64% (80% x 80% = 64%) of William's. So, when William is at 64m when William finishes. 100m - 64m = 36m, thus William beats Yolanda by 36m.

Question 44: A

This question can be solved quickly if you first realise that there is no need to calculate both volumes and subtract the larger from the smaller, instead only convert the television dimensions into metres and then calculate 60% of that.

Question 45: E

From the information provided all the flaws listed are valid since David's main point is that he has chosen the cheapest. A could be true as there is an additional cost of £3 for staying at Whitmore, therefore if the vehicle they are using achieves sufficient miles per gallon then travelling the extra few miles could cost less than £3 in terms of petrol. B again is possible which would argue against it being cheap, as would D. And if C is true then David's argument is flawed altogether.

Question 46: D

C is irrelevant as nowhere does the passage mention standards of modern medical practice. A may be incorrect as nowhere does the article explicitly say that animal testing is the only accepted method of drug approval. B categorically conflicts with the first sentence of the second paragraph.

Question 47: A

Begin by converting all the quantities into terms of items as that is the terminology used on the graph axis. Therefore 12 rugby balls = 6 items and 120 tennis balls = 24 items. Reading from the graph reveals their respective prices as £9 and £5. Therefore, the total cost of products in the order is (6 x 9) + (24 x 5) = 174. Since this is significantly more than £100 the delivery charge is waived.

Question 48: D

Calculate the cost of 10 of everything as (2 x 5) x (10 x 7) x (5 x 9) = £125. Recall that delivery charge is waived at £100 and this therefore a trick question and no delivery charge is applied anyway.

Question 49: E

Tennis balls are sold in the largest pack and so they must be considered. Begin by dividing 1000/5 using the value from the first column = 200. As this is above the range 0 -99 look up the item value in the 100 -499 range where a £1 discount is applied per item. Therefore, in actually fact 1000/4 = 250 items can be purchased which equates to a total of 250 x 5 = 1250 balls.

Question 50: B

Recognise that 120% profit is equivalent to 220% of the original price. In which case the initial purchase price = (1,320/220) x 100 = £600.

END OF SECTION

Mock Paper E: Section 2

1. *"Nature over nurture"*. **To what extent do you agree with this statement?**

Introduction:
- Define briefly what 'nature' and 'nurture' entail – for example, nature can refer to natural attributes such as hereditary traits (e.g. IQ, height, eye colour) whilst nurture can refer to one's environment and circumstances
- State your standpoint as to why you believe nature trumps nurture or vice versa
- Example standpoint: I believe that nurture trumps nature as successful people tend to be brought up in favourable conditions and are surrounded by encouraging and inspirational people. These conditions all constitute 'nurture' and play a larger role in determining one's success as opposed to what natural traits the person was born with (e.g. looks, intelligence)
- Example alternative standpoint: 'Nature over nurture' is largely true due to the fact that even though one's circumstances play a role in determining one's success, ultimately one's genetic traits play a disproportionately larger role in determining a person's pathway in life.

Possible arguments for nurture over nature:
- Successful people tend to be brought up in favourable environments and are surrounded by motivational people
- For example, many successful musicians tend to be brought up in a family tend encourages the study of music, or had parents who were successful musicians themselves
- It is well-known that more well-to-do families tend to produce more well-adjusted children who have a statistically higher chance of earning more in their working life
- Peer influence plays a huge role in the development of a child – e.g. if a child hangs out with children who constantly skip school and engage in gang violence, the child is most likely going to be negatively influenced
- Talent is not enough without hard work – successful artists, musicians and scientists had to work very hard to get to where they are even if they had some talent to begin with
- Studies have shown that a person's characteristics are determined by a combination of genetic factors as well as environmental factors – the study of epigenetics shows that environment plays a huge part in determining what traits a person develops

Possible arguments for nature over nurture:
- Successful people like Einstein, Bill Gates, Tim Cook etc. were all naturally brilliant and clearly had the intellectual capacity to achieve great things, and intellectual capacity is very much a hereditary trait
- Successful actors and models are largely successful due to their good looks, again another hereditary trait
- Studies have shown that good-looking people tend to progress faster at work as they are naturally more confident and are better at networking and establishing good client relationships
- Having more natural intelligence also means that these people can succeed in life much easier with less hard work
- If you are born with a serious defect, or if you were born with deformities, you are naturally going to be much more hindered in life compared to luckier individuals who were born with no conditions
- Even though genetic deficiencies and conditions can be overcome, it does make one's life harder and the starting point in life is drawn largely based on one's luck in the genetic lottery
- Studies have shown that psychopathic criminals tend to have deficiencies in certain chemicals that result in a different expression of mood in their brains, which shows that serious criminals are not produced based solely on their circumstances

Conclusion:
- Include a summary of all points given and refer back to your original definition and standpoint
- You do not have to come up with an extreme conclusion (e.g. nature always trumps nurture, or vice versa), a sensible conclusion that is measured and discusses the possible exceptions is perfectly acceptable

2.   *Do you think positive discrimination should be implemented in our university admission process?*

Introduction:

- Define positive discrimination – the act of lowering entry requirements for certain minority groups in order to overcome inequality and disproportionate representation and improve social mobility
- You may give refer to particular examples such as the US implementing this in their university admission process
- State your standpoint – whether you agree that this should be implemented in order to improve the representation of minority groups in our universities, or whether this goes against meritocracy and produces more harm than good

Possible arguments for positive discrimination:

- Certain minority groups such as black students come from less well-to-do families and do not have access to the same resources in order to do well for their examinations
- Minority students are less encouraged to apply for top universities because they perceive these universities as elite and lacking representation of minority groups
- Minority students do not receive adequate support and consideration for differences in their background and community
- Diversity in university education is important in order to foster a more vibrant community
- It is important to allow more disadvantaged minority students to enter top universities if they have the intellectual capability as this will improve social mobility by improving their chances of securing top-paying jobs upon graduation
- Many minority students are given a lower starting point in life due to inherent discrimination and stereotypes in society, as well as their less well-to-do backgrounds, hence positive discrimination is needed in order to level the playing field
- Universities benefit from having a larger representation of minority students amongst their student population as they bring in different ideas, perspectives and thoughts, which stimulate intellectual academic discussion and studies
- Studies have shown that diversity leads to greater productivity and fosters creative thinking and innovative solutions
- Admissions should not be based solely on raw grades as this does not take into account the fact that many minority students might not have the same resources or teaching in order to score as well as they are capable of in the examinations

Possible arguments against positive discrimination:

- We should not lower entry requirements for a particular group just because they happen to be a minority group as this goes against the idea of admissions based solely on merit
- People might start to have the perception that minority students only managed to enter some top universities due to their race – a situation that is happening in the US
- Majority groups may feel that it is unfair that they failed to get into a university compared to a minority student even though they have the same grades just because of their race
- Lowering entry standards may result in some students entering universities they are not suitable for academically
- Positive discrimination is a slippery slope and might lead to identity politics – whereby students become increasingly polarised and segregated based on their identity
- Many of our top universities pride themselves in being meritocratic and only accepting the best of the best – arguably lowering entry requirements just for a particular group of students risk harming our high standards
- Instead of implementing positive discrimination, we should instead focus more on tackling the problem of social inequality and social mobility by ensuring that minority students receive the education they deserve at the lower levels

Conclusion:

- Include a summary of all points made, and give a balanced overview of both sides of the argument.
- Reiterate what your overall conclusion is and why you have arrived at that conclusion
- A sensible and measured conclusion is desirable – if you have a strong viewpoint, be prepared to defend it

3   *'Developed countries have a responsibility to help developing countries grow'.*
      Do you agree with this statement?

Introduction:

- Define the difference between developed and developing countries
- Give some examples – e.g. developed modern countries such as the US and Japan vs. developing countries that require infrastructural support such as certain Southeast Asian countries and African countries
- State your standpoint – why should developed countries help developing countries grow, or why not?

Possible arguments for:

- Many developing countries managed to develop at a fast pace by exploiting the resources from developing countries
- E.g. Colonialism was known to have helped many countries such as the UK, Spain and Portugal develop quickly at the expensive of the colonised countries
- It is also in the interest of developed countries to help developing countries grow due to the interconnectedness of economies these days
- With greater economic growth, developing countries will experience less instability issues and this will result in greater world peace, which is of benefit to developed countries
- Developed countries have greater resources and have a responsibility of sharing these resources with developing countries in order to help them develop and lift their population out of poverty
- Economic growth is not a zero-sum game and it does not mean that if developed countries help out developing countries, they reduce the amount of resources available for their own people – on the contrary, helping developing countries grow has tremendous knock-on effect in the long run by improving trade relations and diplomatic relations between the two countries
- Developing countries tend to have an abundance of natural resources and developed countries stand to benefit from such resources if they are willing to help developing countries with their infrastructural growth
- In terms of a humanitarian standpoint, many developing countries suffer from acute problems such as lack of clean water, food and resources, and developed countries have a humanitarian duty to do their best in helping these countries and ensuring that people have access to clean water and sanitary conditions
- Having developed countries help developing countries also reduces the risk of inter-country conflict and war as they become more dependent on each other

Possible arguments against:

- Developed countries have their own problems to take care of (e.g. slowing economic growth, strained infrastructure, congestion problems etc.)
- Developed countries may not know how to help developing countries due to the intricacies of their problems
- Sending aid to developing countries does not always work – e.g. sometimes developing countries become overly-dependent on the aid
- Developing countries need to be more independent in developing on their own pace and manner in order for them to be sustainable in the long run
- There have been situations where developed countries have made developing countries worse off by attempting to help them – for example causing more pollution and damage, or stagnating the developing countries' growth by making them heavily-reliant on aid
- Resources are scarce and developed countries can only do so much to help developing countries that are miles away without compromising on the attention they need to give to their own citizens
- Developing countries need to have more effective governance in order to lift their own people out of poverty – this is something that developed countries have little control over due to sovereignty issues
- If the government of a developing country is heavily-corrupted, or if the country is highly sealed off from the world and insular, there is not much developed countries can do to help the country without risking heightened conflict

Conclusion:

- Include a summary of all points made, and give a balanced overview of both sides of the argument.
- Reiterate your overall conclusion and explain how you arrived at your conclusion
- Provide a sensible and measured overall conclusion based on the arguments and counter-arguments you have raised

4   Should our education system place a greater emphasis on the sciences as opposed to the arts?

Introduction:

- Define both the sciences and arts and explain how an education in both might differ
- Contrast the different subjects considered – e.g. learning about physics, biology and chemistry in the sciences vs. learning about art, history and literature in the arts
- State your standpoint – should more emphasis be given to the sciences or should the arts not be neglected?

Some points as to why science should be given a greater emphasis:

- Science is experimental – involving accurate and detailed study.
- Science could be argued to be physical and measurable- using known concepts in the real world that can be objectively identified.
- Science is adaptable and ever changing – according to research and current thought.
- Science could be considered to be held within the thoughts and brains of society – if there weren't people exploring it or interested in it, would it really exist?
- Science involves logical and critical thought – in that sense it could be considered to be merely a thought process – but you may wish to explore the idea that it couldn't exist without physical things to measure in themselves.
- Valuable for the evolution of society – as we learn more about the world and about ourselves, this helps change to occur which is vital.
- Provides concrete fact and stability.
- Gives a purpose to life, which brings fulfillment and contentedness.
- Brings explanation for some things that can't be explained through simple observations, which allows a firmer foundation and an answer to bigger, crucial questions.
- Helps us to gain more of an understanding of the world we are in – bringing a stronger grasp on reality and greater knowledge and insight.
- Medical reasons – drugs, medicines, health.
- Safety – e.g. monitoring volcanoes, predicting earthquakes etc.

Some points as to why arts should not be neglected:
- Learning the arts teaches us how to express ourselves
- Learning the arts foster creative thinking
- Arts place an emphasis on critical thinking
- Learning about history and philosophy can be useful in shaping political and critical thought
- Effective writing and communication is essential and can be picked up from the arts
- Arts comprise a wide variety of subjects – ranging from literature to history to music – all these subjects inculcate different skills and many students have differing levels of aptitude and interest that can be catered by the different subjects
- Many successful politicians and government officials studied arts subjects – clearly being good at an arts subject can be indicative of certain useful attributes such as being an effective communicator, being a critical thinker and being able to see the wider picture
- While the sciences are important, learning the arts help to maintain and develop our culture and helps students develop an appreciation of the finer things in life
- Languages are also increasingly important these days within our globalised world – whereby multilingual individuals are highly favoured in the workforce and a premium is available
- In this day and age, technical abilities are not the only thing employers look out for – communication and effective writing skills are also highly important and these are adequately trained by arts subjects

Conclusion:

- Summarize the key ideas explored previously – the ideas associated with science and arts and their relative importance
- Ensure to link back to the original statement – should more emphasis be placed on the sciences in our education system?
- Provide a good summary of the arguments and counter-arguments you have raised and why you have arrived at your overall conclusion

Mock Paper F: Section 1

Question 1: B

Lucy must live between Vicky and Shannon. Lucy is Vicky's neighbour, so Shannon cannot have a red door. Vicky lives next to someone with a red door, so Lucy must have the red door. This leaves Shannon with the blue door and Lucy with the white. The green door is across the road and so does not belong to any of them.

Question 2: E

First calculate an average complete one-way journey time as 40 + 5 + 5 = 50 minutes. Deducting his breaks, he works a total of 7 hours 20 or 440 minutes. Since the first train is already loaded his first run will only take 45 minutes leaving 395 minutes to complete his working day. 395/50 = 7 remainder 45. Note that 45 minutes is not enough to fully unload the train, but it is enough to load the train and drive the distance. Therefore, the driver will complete a total of 9 journeys equalling a distance of 198 miles.

Question 3: E

A is not actually a valid assumption as we do not know what proposal conservationists might be bringing to the local councils, they have only expressed their concern. They may well be bringing a proposal to ask for funding to rehome all the species in the affected environment. B is essential to the final paragraph whilst C must be assumed otherwise the councils would not be presenting these proposals at all.

Question 4: D

As there is not really information in the question to calculate the answer quickly. Instead consider each answer in term and calculate the differences to find the correct price difference in the question:

    A) (3x 1) + (2 x 1.25) – (15 x 0.3) + (10 x 0.5) etc…

Question 5: E

Based on the information in the question options A - D are simply wrong. A is incorrect as antibiotic E has not affected growth at all. B is incorrect as the other antibiotics have significantly affected growth. E was the least effective antibiotic. C was not the most ineffective as it did disrupt growth slightly whereas E had no effect at all. D will now be taken and the experiment repeated with D at numerous concentrations to find the optimum dose.

Question 6: A

1L = 1000 cubic centimetres and therefore the total volume of air Laura needs to produce is 25 x 0.3 = 7.5L. With a total of 25 balloons she will take 25 x 0.5 = 12.5 seconds breathing in and a further 7.5/4.5 = $1\frac{2}{3}$ minutes inflating the balloons. This yields a total time of 1 minute 40 + 12.5 seconds = 1 minute 52.5 seconds or 112.5 seconds.

Question 7: B

Quickly represent the question schematically as (A = B) ≠ (C = D = E). We can now observe that A in fact supports George's argument, C also supports George's argument and D may well be true, but it would have no effect in disrupting the argument, simply only imply D and E are both also equal to 0. However, as E is equal to C it should therefore not equal B.

Question 8: B

This question is much less complicated than it sounds. Begin by just considering a single hour. Throughout the hour of 1 the hour hand will be pointing at 1. Only during the 5th minute of that hour will the minute hand point to the 1 whereas every 5th second of the minute the second-hand points to the 1. All these events will only coincide once. As there are 24 hours in a day 00:00 through to 23:59 this event will happen 24 times.

Question 9: B

A) Potentially correct, but extreme sports also carry higher risks of injury.

B) True.

C) True, but irrelevant for the question.

D) Potentially correct, but irrelevant to the question.

Question 10: D

A) False – we are not told about the healthcare directly but are told that injury and disease posed a threat.

B) False – the terrain was difficult, and mapping was poor.

C) False – outlaws were a significant threat.

D) True – as the text states, there was a marked lack of bridges.

Question 11: C

Whilst B may be true it is not a reason for dependence, only a supporting factor. Dependence implies that we have no choice but to use electricity. Hence A is wrong as gas is readily available; hence D is wrong for the same reason. This leaves the correct answer C which is the only statement which truly describes our absolute necessity for electricity – since electrical appliance by definition only function with electricity.

Question 12: B

First note that 27 guests plus Elin herself means that 28 people will be eating the 3 courses which will require a total of 28 x 3 = 84 glasses of wine. This is a total volume of 84 x 175 = 14.7L = 21 bottles. As wine is only sold in cases of 6, Elin will have to buy 24 bottles so as not to run out. Recall the buy one get one free offer so she only pays for 2 cases.

Question 13: E

Recognise that when rounded to the nearest 10 the shortest an episode could last is 35 minutes. Hence a total of 7 x 12 = 84 episodes would take a total of 84 x 35 = 2,940 minutes = 49 hours.

Question 14: E

Points A and B are the best exemplified through this passage. Often great discoveries come from accidental observations and then exact processes are refined through many experiments in a trial and error fashion until the correct methodology is achieved. The passage demonstrates how as our understanding of the world around us advance so too does our ability to provide healthcare. D can be observed in the passage as the 50/50 split.

Question 15: C

Whilst A and D are true they do not force the stranger to give him the sapphire – remember Jack can be given any stone for a truthful statement. B and C are both lies and will earn Jack nothing. Instead if Jack states E then the stranger has no choice to hand over the sapphire else it would be a false statement.

Question 16: A

Despite the enormous interest rate in Simon's current account it is only awarded twice, whereas in the saver account it is awarded 4 times. Hence earnings from the saver account = 100×1.5^4 = £506 whereas earnings from the current account would have stood at £361.

Question 17: D

The largest possible key can be obtained were the first two numbers are at a maximum because they are multiplied together → 9 x 9 = 81. Subtract the smallest number to yield 81 – 1 = 80 and again divide by the smallest number which is 1 hence 80 is the largest possible key.

Question 18: D

A) Incorrect. The text clearly states that the exercise routine is resistance training based.

B) False. Both groups contain equal numbers of men and women per the text.

C) False. Both groups are age matched in the range of 20 to 25 years.

D) Correct. As the only difference between the two shakes is the protein content.

Question 19: D
There are three different options for staying at the hotel. They could either pay for three single rooms for £180, one single and one double room for £165, or one four-person room for £215.

Subtracting the cleaning cost for one night would leave:
£180-(3x£12) = £144
£165-(2x£12) = £141
£215-£12 = £203

The cheapest option is one single and one double room and they want to stay three nights, giving £141x3 = £423.

Question 20: D
Glass one starts with 16ml squash and 80ml water. Glass two starts with 72ml squash and 24ml water. 48ml is half of 96ml so 8ml squash and 40ml water is transferred to glass two. Glass two now contains (8+72 = 80ml squash) and (24+40 = 64ml water). Glass two now has a total of 144ml and half of this is transferred to glass one. Glass one now has (40+8 = 48ml squash) and (32+40 = 72ml water). Therefore, glass one has 48ml squash and glass two has 40ml squash.

Question 21: B
B is the main conclusion of the argument. Options A and D both contribute reasons to support the main conclusion of the argument that the HPV vaccination should remain in schools. C is a counter argument, which is a reason given in opposition to the main conclusion. Option E represents a general principle behind the main argument.

Question 22: B
The speed of the bus can be calculated using the relationship: Speed $=\frac{distance}{time}$

$\frac{3\ km}{0.2\ h} = 15$ kmh^{-1}

The bike speed is therefore ($\frac{4}{5}$ x 15 = 12 kmh^{-1}). Considering that the bus leaves 2 minutes after the bike, it is now possible to write an expression, where d is the distance travelled when the bus overtakes the bike:

$\frac{d\ km}{12\ km/h} = \frac{1}{30}$ h $+ \frac{d\ km}{15\ km/h}$

This expression can be solved by multiplying each term by (12 kmh^{-1} x 15 kmh^{-1}):
15d km = 6 km +12d km
3d km = 6 km
d = 2 km

Therefore, the bus overtakes the bike after travelling 2 km.

Question 23: B
Firstly, determine who will move up to set one. Terry, Bahara, Lucy and Shiv all have attendance over 95%. Alex, Bahara and Lucy all have an average test mark over 92. Terry, Bahara, Lucy and Shiv all have less than 5% homework handed in late. Therefore, Bahara and Lucy will both move up a set. Secondly, determine who will receive a certificate. Terry, Bahara, Lucy and Shiv have absences below 4%. Alex, Bahara and Lucy have an average test mark over 89. Bahara and Shiv have at least 98% homework handed in on time. Therefore, only Bahara will receive a certificate.

Question 24: C

Firstly, construct two algebraic equations: A-18=B-25 and A=$\frac{5}{6}$B

Solve these two equations as simultaneous equations by substituting $\frac{5}{6}$B for A in equation 1:

$\frac{5}{6}$B-18=B-25

$7=\frac{1}{6}$B

B=42

Put B=42 back into equation 2: A= 42 x $\frac{5}{6}$

A=35

Question 25: D

I need to make 48 scones, which makes up 8 batches.
8 batches would take: 35+ 7(25+10) +25 = 305 minutes

I need to make 32 cupcakes, which makes up 4 batches.
4 batches would take: 15+ (4x20) =95 minutes

I need to make 48 cucumber sandwiches
This would take (8x5) = 40 minutes

Adding 305, 95 and 40 minutes is 440 minutes in total. 440 minutes is equivalent to 7 hours and 20 minutes. Adding 7 hours and 20 minutes to 10:45am leads to 6:05pm so I will be finished at 6:05pm.

Question 26: D

The volume of a pyramid is given by the equation:

$v = \frac{a^2h}{3}$ where v=volume, a=base and h=height

Rearrange to work out the height for each pyramid: $h = \frac{3v}{a^2}$

Pyramid	Base edge (m)	Volume (m³)	Calculation:	Height (m)
1	3	33	$\frac{3x33}{9}$	11
2	4	64	$\frac{3x64}{16}$	12
3	2	8	$\frac{3x8}{4}$	6
4	6	120	$\frac{3x120}{36}$	10
5	2	8	$\frac{3x8}{4}$	6
6	6	120	$\frac{3x120}{36}$	10
7	4	64	$\frac{3x64}{16}$	12

The tallest pyramid is 12m and the smallest is 6m. Subtracting the height of the tallest pyramid from the height of the smallest pyramid leaves 6m.

Question 27: A
Work out the two wages by substituting the information provided into the formula:
Jessica's wage is: 210 + 42 - 3.2 = 248.8
Samira's wage is: 210 + 78 - 8.8 = 279.2

Subtracting 248.8 from 279.2 leave 30.4 so the difference between their wages is £30.4.

Question 28: C
The main conclusion is C. A and B both represent reasons to support the main conclusion of the argument. Option D represents an assumption that is not stated in the argument but is required to support the main conclusion that research universities should strongly support teaching. Option E is a counter argument that provides a reason to oppose the main argument.

Question 29: D
D is the main conclusion of the argument. A is a general principle of the argument, but the argument is more specific to the use of helmets rather than the wider concept of danger in sport and the responsibilities of the governing bodies to sports players. Options B and C are reasons to support the main conclusion. Option E is an intermediate conclusion, which acts as support for the next stage of the argument and as a reason to support the main conclusion.

Question 30: D
There are 10 passengers on the tube at the final stop. At stop 5 there were twice the number of passengers on the tube so 20 passengers were at stop 5. At stop 4, there were $\frac{5}{2}$ times the number of passengers at stop 5 so 50 passengers were present at stop 4. At stop 3, there were $\frac{3}{2}$ times the number of passengers at stop 4 so 75 passengers were on the tube. At stop 2, there were $\frac{6}{5}$ times the number of passengers at stop 3 so 90 passengers were present at stop 2. Similarly, at stop 1, there were $\frac{6}{5}$ times the number of passengers at stop 2 so at the first stop 108 passengers got on the tube.

Question 31: E
➢ Some students born in winter like English, Art and Music
➢ There is not enough information to tell whether some students born in spring like both Biology and Maths.
➢ We don't know what the students born in spring think about Art.
➢ We don't know what the students born in winter think about Biology.
➢ There is not enough information to know whether this is true or not.

Subject	Time of Birth		
	Spring	Autumn	Winter
English	Everyone likes	Everyone likes	Everyone likes
Biology	Some like	No one likes	
Art		Everyone likes	Some like
Music			Everyone likes
Maths	Some like		

Question 32: A
The main conclusion is option A that some works of modern art no longer constitute art. B is not an assumption made by the author as the main conclusion does not rely on *all* modern art being ugly to be valid. C is not an assumption because the argument does not rely on artists studying for decades to produce pieces of work that constitute art. This point is simply used to support the main argument. Options D and E are stated in the argument so are not assumptions. A is an assumption because it is required to be true to support the main conclusion but is not explicitly stated in the argument.

Question 33: E

Reducing the price of the sunglasses by 10% is equivalent to multiplying the price by 0.9. the price of the sunglasses is successively reduced by 10% three times and so the price on Monday is 0.9^3 the price of the sunglasses on Friday. 0.9^3 is equal to 0.729 and so the price of the sunglasses on Monday is 72.9% of the price of the sunglasses on Friday.

Question 34: C

It is easier to write out this calculation in the following format:

```
a b 7 –
  a b
_____

5 6 5
```

From the above subtraction it is clear that b must be equal to 2 because 7 minus 2 is equal to 5, which is the unit term of the answer. It is now possible to rewrite the calculation with 2 substituted for b:

```
a 2 7 –
  a 2
_____

5 6 5
```

From the above calculation it is possible to gauge certain facts. A must be greater than 5 because 1 is carried over to the second term:

```
a ¹2 7 –
   a 2
_____

 5 6 5
```

It is now clear than a must be equal to 6 because 12 minus 6 is equal to 6, which is the tens value of the answer.

Question 35: E

Look at the flat cube net and note the shapes that are adjacent to each other. Sides that are joining on the net will be beside each other on the formed cube. Work through to deduce option E can be formed from the cube net shown.

Question 36: E

The H shape is comprised of 12 squares. The shape's area of 588 can be divided by 12 to give 49, which is the area of each individual square. The square root of 49 is 7 and so the side length of each individual square is 7cm. The perimeter of the shape is comprised of 26 sides and the length of each side is 7 so the perimeter of the shape is 182cm.

Question 37: E

The information provided about the child needs to be inserted into the BMI formula: $BMI = 35 \div 1.2^2$
1.2 squared is equal to 1.44 and it may be easier to work out 3500 divided by 144. The answer needs to be worked out to 3 decimal places for an answer required to 2 decimal places. The answer to 3 decimal places is 24.305 and so the BMI to 2 decimal places is 24.31.

Question 38: C

It is important that the information is inserted into the formula given for calculating the BMR of a woman rather than a man:
BMR= (10 x weight in kg) + (6.25 x height in cm) – (5 x age in years) -161
BMR = (10 x 80) + (6.25 x 170) – (5 x 32) – 161
BMR= 800 + 1062.5 -160 -161

The BMR of the woman in the question is therefore 1541.5 kcal

Question 39: D

This time, the information needs to be inserted into the formula for calculating the BMI of a man:

BMR= (10 x weight in kg) + (6.25 x height in cm) – (5 x age in years) + 5

BMR= (10 x 80) + (6.25 x 170) – (5 x 45) +5

BMR= 800 + 1062.5 -225 +5

The BMR of the man in the question is therefore 1642.5 kcal. The man does little to no exercise each week. It is therefore required to multiply 1642.5 by 1.2, which gives a daily recommended intake of 1971 kcal.

Question 40: B

Slippery Slope describes a series of loosely connected and increasingly worse events that lead to an extreme conclusion. A is not a flaw because the author does not predict a series of undesirable outcomes. C is not a flaw. It is unlikely that correlation has been confused with cause if the American school did not change other aspects of the school day although this is not explicitly stated in the argument. D is not a flaw. A circular argument assumes what it attempts to prove and this is not the case in this argument. E is a counter argument rather than a flaw. B is the flaw in the argument. Just because moving start times later worked in one school in America does not mean that it will work in all other cases.

Question 41: D

Options A and E, if true, would weaken the argument. If the class is more disrupted this will be detrimental to learning, as will less effective teaching. B does not strengthen the main conclusion, which is based on improvement in academic achievement levels rather than activity levels. C does not strengthen the argument as the school curriculum makes no difference to the argument about the science behind teenage brains. If D is true then it suggests that the improvement in grades is a direct effect of the later school starts rather than a mere correlation.

Question 42: B

The main conclusion is that EnergyFirst is expected to expand its customer base at a rate exceeding its competitors in the ensuing months. A does not directly contradict the main argument. It demonstrates a flaw in the argument in that it ignores the fact that other companies may be stronger in other areas and attract customers by other means. However, it does not serve to weaken the main argument. C does not contradict the main conclusion; EnergyFirst could still expand its customer base at the fastest rate even if there is not much competition between energy companies. D would not weaken the argument as it refers to the rate of new customer intake rather than the number of new customers attracted. E, if true, would strengthen the argument because it suggests that visual advertising would attract new customers. B would weaken the main argument because if it were true then investing the most money in advertising would not serve to attract the most customers.

Question 43: A

Option B points out a flaw in the argument, which attributes the healthier circulatory system of vegetarians to diet, but ignores other potential contributory factors to a healthy circulatory system such as exercise. C is not an assumption: the health benefits of a vegetarian and omnivorous diet are not discussed; rather the argument is centred on the negative health ramifications. D is stated in the argument so is not an assumption and option E is a counter argument, not an assumption. Option A is required to support the main conclusion but is not stated in the argument so is an assumption made in the argument.

Question 44: A

First, calculate the number of hours spent flying and waiting. It takes 24 hours in total from Auckland to London, 11.5 hours from London to Calgary and 8 hours from Calgary to Boston. In total this amounts to 43.5 hours of flying and waiting. Boston is 16 hours behind Auckland and so when Sam arrives in Boston it will be 27.5 hours ahead of 10am. The time in Boston will therefore be 13:30 pm.

Question 45: B

This question requires you to find the lowest common multiple. This is the product of the highest power in each prime factor category.

$18 = 3^2 \times 2$ $33 = 3 \times 11$ $27 = 3^3$

Therefore, 3^3, 11 and 2 need to be multiplied together which equals 594 seconds between simultaneous flashes. 5 minutes or 300 seconds needs to be subtracted from 594 in order to find the length of time until the next flash. The time that they will next flash simultaneously is 294 seconds.

Question 46: B

Firstly, calculate the number of students who play each instrument. 21 students play piano, 12 play violin and 3 play saxophone. Point 1 is true because the sum of 21 piano students and 12 violin students is 33, which is 3 more than the total number of students in the class. Therefore, at least 3 students must play both piano and violin. Point 2 is true because only 12 students actually play the violin so there cannot be more than 12 students playing both piano and violin. Point 3 does not have to be true because some of the 9 students that do not play piano may play the violin.

Question 47: D

One way of answering this question is to set out the result after each game:

	Neil	Simon	Lucy
Start	50	50	50
Game 1	100	25	25
Game 2	50	50	50
Game 3	25	25	100
Game 4	12.5	12.5	125
Game 5	6.25	15.625	128.125

After game 5 Lucy has £128.13, however the question is asking how much money Lucy gains. The difference between 128.125 and 50 is 78.13, so Lucy gains £78.13.

Question 48: C

Option A may explain why young drivers are involved in more accidents but does not need to be true for the main conclusion to hold. B would weaken the argument if true as drivers that spend more time driving will have a greater chance of being involved in accidents regardless of age. D is not an assumption, but if true may weaken the argument as it attributes the accidents to unsafe cars rather than unsafe driving. E is irrelevant to the main conclusion: it does not matter whether the young drivers are male or female; arguably steps should still be taken to reduce the number of accidents. Option C represents an assumption that is not stated in the argument but is required to support the main conclusion.

Question 49: D

The total weight of all of the apples is 6 multiplied by 180g, which equals 1080g. The highest value the heaviest apple could take would occur if all of the other 5 apples weighed the same as the lightest apple. 5 multiplied by 167g, the weight of the lightest apple, is 835g. The difference between the weight of all of the apples (1080g) and 835g gives the highest possible weight of the heaviest apple, which is 245g.

Question 50: B

A) True, but not far-reaching enough.

B) Correct answer. Sugar does indeed have an addictive potential as it causes the release of endorphins and the health concerns are well known. This characteristic makes it like alcohol and smoking, and potentially suitable for similar policies.

C) True, but similar to option A) and thus too limited.

D) Potentially true, but also too limited.

Mock Paper F: Section 2

1   'Social media has risen and fallen'. Discuss.

Introduction:
- Define briefly what social media entails – e.g. Instagram, Facebook, Twitter, Snapchat, YouTube
- State what can possibly be attributed to the 'rise' and the 'fall' of social media
- Indicate what your overall standpoint is, for example, state why you think social media has risen over the years (increasing number of users, financial success of these social media companies, social media becoming quintessential in our lives) and why there might have been a recent fall (e.g. people eschewing social media platforms, drop in interest over social media platforms)

Rise of social media:
- Exponential increase in the number of social media users over the past ten years
- Social media has become highly influential – e.g. advertisers regularly use social media platforms as a tool to reach their target audience
- Social media platforms have been used for philanthropic and humanitarian purposes – e.g. spreading awareness of certain diseases, plights etc.
- Social media companies have gained significant international clout – e.g. Facebook and Snapchat IPOs
- Social media has becoming a quintessential part of our lives – not only are the younger generation using it, nowadays the older generation have also caught up with the trend and many of them have been using Facebook as a platform to connect with their old friends, colleagues, family members and so on
- There seems to be a platform catered for differing interests and individuals – e.g. Facebook is largely used for staying in touch with friends and relatives, Instagram is for the younger generation and is usually filled with pretty pictures (e.g. holiday pictures, fitness posts, food pictures), whilst Snapchat is predominantly used by teenagers who prefer to chat with each other with the benefit of more privacy
- Social media has also become a lucrative platform – for example certain YouTube stars are able to rake in millions just by posting seemingly regular and mundane videos such as videos of themselves playing games or just chatting about life in general
- Top Instagram stars who have millions of followers are also able to demand premium prices for displaying sponsored products in their posts
- Twitter has also become somewhat of a political tool – as seen from certain politicians such as Donald Trump who uses Twitter frequently as a tool to publish his unbarred thoughts

Fall of social media:
- Increasing criticism over fake news – e.g. Facebook failing to filter out fake news from its News Feed
- Declining relevance of certain platforms such as Snapchat – e.g. fall in share price
- Lack of security – e.g. social media being used by paedophiles to target vulnerable children
- There has been a rapid decline in users in certain platforms – e.g. Facebook has been losing a lot of young users as they perceive Facebook to be increasingly flooded with the older generation and they feel a loss of privacy
- There is a certain level of disenchantment with social media usage – e.g. more and more people are boycotting social media over the fact that it leads to a loss of privacy, creates an unhealthy obsession with portraying a certain image, and it results in a popularity contest at times
- Social media companies have also been criticised for failing to treat the privacy of individuals more seriously – e.g. the data accumulated from a user's social media usage has been used to created targeted advertisements, and such data can be potentially intrusive and include details such as a user's address, ethnicity, sexual orientation, health matters etc.
- Young girls in particular have become more susceptible to eating disorders as a result of the unhealthy body image promoted by social media usage

Conclusion:
- Include a summary of all points given and refer back to your original definition and standpoint

2   Do you think the right to press freedom has become increasingly dangerous and polarising to society?

Introduction:
- Briefly state the notion of the right to press freedom and why it is protected
- Explain why this right might have become increasingly dangerous and polarising to society
- State your standpoint
- For example, you should mention why right to press freedom has traditionally been protected – e.g. to preserve journalistic integrity and to ensure that the people have a right to know about what is going on (such as government corruption)
- You should then go on to talk about how this right might have manifested over the years and has been abused – for example irresponsible and unscrupulous journalists who create sensationalist and fake news articles in order to generate more readership

Possible reasons for why it has become increasingly dangerous and polarising:
- Lack of effective quality check – e.g. rise of fake news
- Irresponsible journalism leading to the public being misled – e.g. bad reporting during the Brexit campaign
- Sensationalist articles polarising society – e.g. articles bashing migrants
- Far-right activists inciting hate by delivering controversial speeches
- Used as a tool for slandering and defamation of politicians
- Right-wing news articles such as The Sun and The Daily Mail are particular notorious for focusing on gossip and publishing news articles in a way that will increase their readership – e.g. sensationalist headlines, focusing on stereotypes, pandering to the right-wing community by publishing opinions that come across as intolerant and insensitive
- On the flip side of the coin, left-wing newspapers have also been known to public news articles that are heavily-skewed and polarising – e.g. The Guardian publishing news articles that are very liberal and insistent on certain topics which might not be agreeable to a more conservative group of people
- Journalists have been heavily squeezed by their profit margin and they can only survive by increasing their readership, and may have to resort to focusing on sensationalist news even if it goes against their professional integrity – e.g. The Sun and The Daily Mail have a much higher readership than considerably more neutral or high-brow newspapers such as The Times or The Telegraph
- People choose to read what aligns with their views so a reader will only read right-wing newspapers and reinforce their pre-conceived notions, and the same goes for left-wing readers who will only actively read left-wing newspapers – this results in society becoming increasingly polarised and people on both sides of the spectrum become more extreme and intolerant of alternative viewpoints

Possible arguments against:
- Press freedom crucial in ensuring an effective check and balance against the misuse of power
- Crucial in exposing scandals
- Press freedom has been pivotal in the recent Harvey Weinstein scandal and the #MeToo movement
- Public has a right to unhindered access of information in order to form their own informed decision
- There is a plethora of alternative news articles available in the market – instead of restricting press freedom we should encourage people to read up more widely and read from different sources in order to come to a more sensible and informed viewpoint
- A free and independent journalism industry is key to ensuring that we are a functioning and effective democracy by ensuring that the population is always kept informed about what is happening in the country and in the world, and bad policies and scandals cannot be swept under the rug easily
- Without press freedom, many controversial incidents would not have come to light, such as the Panama and Paradise Papers

Conclusion:
- Include a summary of all points made, and give a balanced overview of both sides of the argument.
- Reiterate what your overall conclusion is and why you have arrived at that conclusion
- As always, if you have a strong opinion, make sure you defend it well and state in the conclusion why the arguments or counter-arguments you have raised made you come to a certain conclusion

3 *'Artificial intelligence replaces humans and leads to unemployment'. Do you agree with this statement?*

Introduction:
- Define artificial intelligence
- Give possible examples of the use of artificial intelligence – e.g. driverless cars, automation
- Explain why you think this might or might not lead to unemployment
- For example, you can mention how artificial intelligence has already been slowly phased out in certain industries – e.g. replacing call centres, the use of artificial intelligence in smartphones and even self-checkout machines have been replacing cashiers

Possible arguments for artificial intelligence:
- Artificial intelligence supports the economy rather than compromising it
- Artificial intelligence leads to greater efficiency
- Artificial intelligence reduces human error
- Artificial intelligence creates more high-end jobs that require more skills such as coding and engineering
- Mundane jobs that are low-paid and menial will be replaced with high quality jobs
- Artificial intelligence is highly complementary to human labour as it reduces the need for humans to do repetitive and menial chores and allows workers to focus on more complex and sophisticated manners
- Artificial intelligence has a wide application and can be used for anything such as self-checkout machines to being employed in complex, legal transactions by automating certain processes
- The rise of artificial intelligence has led to the boom of our tech industry and has created many relevant jobs in the field of software engineering and coding
- Artificial intelligence has solved many problems such as a declining population in certain countries such as Japan which necessitates artificial intelligence in some traditionally labour-intensive industries
- Artificial intelligence has also made it possible for the economy to be more efficient, such as allowing driverless trains to operate and hence allowing for a 24-hour service
- Artificial intelligence has also been used in high-end medical technology due to its much greater precision, helping to improve healthcare

Possible arguments against artificial intelligence:
- Artificial intelligence disproportionately affects low-income groups as their jobs are the most susceptible to automation
- Artificial intelligence might result in a lack of human touch – e.g. replacing drivers and waiters
- Artificial intelligence is potentially unreliable and unstable
- Certain jobs are not amenable to artificial intelligence – e.g. jobs requiring communication and interaction
- The cliché of artificial intelligence creating super robots that take over humankind may have some truth to it as robots become more sophisticated and are able to mimic human actions more – there is a limit as to where our ethical boundaries are and to what extent should we make such robots humanlike
- Whilst artificial intelligence is promising and seems to be able to develop our economy, it is still largely uncharted territory and it is questionable whether artificial intelligence can truly replace certain skilled jobs such as lawyers and doctors which require years of training
- Automation results in a massive cut in labour-intensive jobs such as manufacturing and production – these workers may not always be able to upgrade their skills to keep up with the changing times and they may become unemployed as a result
- People are resistant to change and certain group of workers may be heavily protected by trade unions and refuse to lose their jobs to automation – e.g. train drivers

Conclusion:
- Include a summary of all points made, and give a balanced overview of both sides of the argument.
- Reiterate your overall conclusion and explain how you arrived at your conclusion
- Come up with a sensible conclusion based on the arguments and counter-arguments you have raised, and briefly explain why you have arrived at this conclusion

4   Does our society place too much emphasis on a university education?

Introduction:

- Briefly state why a university education is emphasised in our society
- State the pros and cons of university education
- State your standpoint as to whether we truly over-emphasise university education
- For example, you can state how many students enrol in subpar universities that do not truly improve their job prospects or do not end up learning anything useful in university due to the lack of proper teaching
- The example of the Blair administration wanting to raise the number of university graduates to 50% of the cohort can also be used to show how this might result in an unsustainable number of university graduates in the job market, leading to a mismatch in demand and supply of certain jobs

Arguments for university education:

- A higher level of education is beneficial to society as a whole
- Keeping students in education will prevent them from being bored and going astray
- University education inculcates useful skills such as teamwork and communication
- University is a good period of time for students to form lasting bonds and networks
- University education is an important sector of our economy
- There is a positive correlation between higher levels of education and future earning power
- University is not solely about academics, even though it forms a huge factor, but there are also several extra-curricular activities that students tend to engage in, such as sports and clubs, which help to develop their leadership and organisational skills
- Arguably being in higher education is a good way of giving students time to properly think about what they want to do in life later on, giving them a safe space to develop their own perspectives and thoughts without the consequences of the real world, and helping them to mature as adults upon graduation
- Being in university also means that you will be part of a network of alumni, which is highly beneficial in the working world when you establish professional relationships and being part of an alumni network is useful for forging connections
- Certain important skills such as effective writing, communication, presentational skills, computer skills and editing skills can be picked up at university which is highly beneficial for working life
- Not all students go into university purely for the sake of getting a good job – some might genuinely enjoy the intellectual stimulation, the friendships that can be made, living in a hall together, or simply if they are interested in academia

Arguments against university education:

- Not all students are academically suitable for university education
- Some universities do not provide a high quality of education
- Many students incur enormous debts from university and are unable to secure well-paying jobs
- Some students waste their time in universities by not focusing on their studies and spend their time partying and drinking instead
- Universities do not always produce graduates that are suitable for the demands of the workforce
- University education has become a huge burden on taxpayers – many student loans do not get paid off fully because many students do not end up with jobs that are well-paying upon graduation
- Universities have also led to a segregation in certain cities – for example the town vs. gown culture in Oxford and Cambridge, and also the increasing gentrification as students dominate certain towns
- Teaching in certain universities have been known to be subpar, and this affects even certain top universities, as professors prefer to focus on their research at the expense of providing high quality teaching to undergraduate students – this is partly due to the fact that university rankings place an emphasis on research quality over student satisfaction
- Universities have become less selective in accepting students and have a tendency to accept a large number of students because they are fee-paying and this results in a larger class size

Conclusion:

- Summarize the key ideas explored previously – the ideas associated with university education and whether we should decrease our emphasis on it
- Ensure to link back to the original statement – should less emphasis be placed on university education?

Final Advice

Arrive well rested, well fed and well hydrated

The TSA is an intensive test, so make sure you're ready for it. Ensure you get a good night's sleep before the exam (there is little point cramming) and don't miss breakfast. If you're taking water into the exam then make sure you've been to the toilet before so you don't have to leave during the exam. Make sure you're well rested and fed in order to be at your best!

Move on

If you're struggling, move on. Every question has equal weighting and there is no negative marking. In the time it takes to answer on hard question, you could gain three times the marks by answering the easier ones. Be smart to score points- especially in section 2 where some questions are far easier than others.

Make Notes on your Essay

You may be asked questions on your essay at the interview which might take place upto six weeks after the test. This means that you **MUST** make short notes on the essay title and your main arguments after you finish the exam. This is especially important if you're applying to Economics or PPE where the essay is discussed more frequently.

Afterword

Remember that the route to a high score is your approach and practice. Don't fall into the trap that *"you can't prepare for the TSA"*– this could not be further from the truth. With knowledge of the test, some useful time-saving techniques and plenty of practice you can dramatically boost your score.

Work hard, never give up and do yourself justice.

Good luck!

Acknowledgements

I would like to express my sincerest thanks to the many people who helped make this book possible, especially the 15 Oxbridge Tutors who shared their expertise in compiling the huge number of questions and answers.

Rohan

About UniAdmissions

UniAdmissions is an educational consultancy that specialises in supporting **applications to Medical School and to Oxbridge**.

Every year, we work with hundreds of applicants and schools across the UK. From free resources to our *Ultimate Guide Books* and from intensive courses to bespoke individual tuition – with a team of **300 Expert Tutors** and a proven track record, it's easy to see why UniAdmissions is the **UK's number one admissions company**.

To find out more about our support like intensive **TSA courses** and **TSA tuition** check out www.uniadmissions.co.uk/tsa

Your Free Book

Thanks for purchasing this Ultimate Guide Book. Readers like you have the power to make or break a book – hopefully you found this one useful and informative. If you have time, *UniAdmissions* would love to hear about your experiences with this book.

As thanks for your time we'll send you another ebook from our Ultimate Guide series absolutely <u>FREE</u>!

How to Redeem Your Free Ebook in 3 Easy Steps

1) Find the book you have either on your Amazon purchase history or your email receipt to help find the book on Amazon.

2) On the product page at the Customer Reviews area, click on 'Write a customer review'

Write your review and post it! Copy the review page or take a screen shot of the review you have left.

3) Head over to www.uniadmissions.co.uk/free-book and select your chosen free ebook! You can choose from:
- ✓ TSA Mock Papers
- ✓ TSA Past Paper Solutions
- ✓ The Ultimate Oxbridge Interview Guide
- ✓ The Ultimate UCAS Personal Statement Guide
- ✓ The Ultimate TSA Guide – 300 Practice Questions

Your ebook will then be emailed to you – it's as simple as that!

Alternatively, you can buy all the above titles at **www.uniadmissions.co.uk/our-books**

TSA Online Course

If you're looking to improve your TSA score in a short space of time, our **TSA Online Course** is perfect for you. The TSA Online Course offers all the content of a traditional course in a single easy-to-use online package- available instantly after checkout. The online videos are just like the classroom course, ready to watch and re-watch at home or on the go and all with our expert Oxbridge tuition and advice.

You'll get full access to all of our TSA resources including:

- ✓ Copy of our acclaimed book "The Ultimate TSA Guide"
- ✓ Full access to extensive TSA online resources including:
- ✓ 20 hours of TSA on-demand lectures
- ✓ 6 complete mock papers
- ✓ 300 practice questions
- ✓ Fully worked solutions for all TSA past papers since 2008
- ✓ Ongoing Tutor Support until Test date – never be alone again.

The course is normally £99 but you can get **£ 20 off** by using the code *"UAONLINE20"* at checkout.

https://www.uniadmissions.co.uk/product/tsa-online-course/

£20 VOUCHER:
UAONLINE20

Made in the USA
Coppell, TX
25 May 2020